Equator

Worldwide distribution of temperature

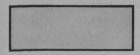

always mild

hot summer
mild winter

predominantly
hot

Collins Guide to

The Weather

Collins Guide to

The Weather

Gunter D. Roth

Translated by

E. M. Yates

Collins
St James's Place, London

William Collins Sons & Co Ltd
London · Glasgow · Sydney · Auckland
Toronto · Johannesburg

ISBN 0 00 219010 9
Wetterkunde für alle was first published in Germany in 1979 by
BLV Verlagsgesellschaft mbH, Munchen
© BLV Verlagsgesellschaft mbH, Munchen 1979
© in the English translation William Collins Sons and Co Ltd 1981
Filmset by Jolly & Barber Ltd, Rugby
Printed and bound by Wm. Collins Sons & Co Ltd, Glasgow

Contents

The weather concerns us all

The only thing certain about the weather is its uncertainty. We have all wondered, on going to work in the morning, whether or not to take an umbrella – despite assurances from the weather forecaster on the radio of a dry sunny day. We have all worried, after carefully booking our holidays for the anticipated hot weather spell in some sunny clime, if this is going to be the one year when the weather will turn as soon as we get there. Short- or long-term forecasts, we fear the worst.

It is this that makes the weather so interesting to everyone. Many of us, owing to the nature of our jobs, are dependent on it. Farmers, fishermen, sailors and pilots, among others, must watch carefully for the slightest change in conditions, for their very livelihoods may be at risk should they fail to observe the signs of hazardous weather. People in communications and in sport, or for that matter any of us who pursue an outdoor leisure activity, are all to some extent dependent on the weather. It is not surprising, therefore, that the weather is often the topic of conversation. How many times have we all discussed or overheard comments on the weather when meeting a friend or acquaintance, or when speaking on the telephone or when writing a letter?

The weather may well govern our moods. A sunshine-filled morning is usually incentive enough to get up and to go out as soon as possible, whereas staying indoors huddled next to a fire often seems the most effective way of overcoming the gloom brought on by rain or snow.

No wonder we mutter about the seeming inadequacies of the weather forecaster who, we sometimes feel, almost wilfully misleads us. In fact, his job is an exceedingly difficult one which a host of scientific instruments, mathematical formulae, weather charts and satellite stations has not made any easier. A better understanding of the study of weather would enable us to interpret his prognoses to suit our own plans and requirements, and we might even venture towards a certain amount of forecasting ourselves. This understanding, then, is the principal aim of the book.

Britain, together with the rest of western and central Europe, has a varied and changeable climate. Some conditions are seasonal, such as the ice, snow and slush of winter and the 'March winds and April showers' of spring, but winter does not always bring snow and it has been known to fall in July. Summer may be a succession of golden days but rain belts frequently occur, even in central Europe, for westerly air streams can predominate and the walker or climber, setting out with clear skies, may be overtaken by a sudden shower. An 'Indian summer' in late September or early October brings an extended period of sunshine, although by late October the days are considerably shorter and the incoming heat from the sun (radiant energy) is much reduced. Clear days give way to fog; autumnal storms at sea and on the coast cause danger to shipping and bring the rains and perhaps even the first snow showers to inland areas.

Although broad patterns of weather do follow the seasons, it is more difficult to explain the rain and low temperatures which occurred in the summer of 1974, that drove dispirited holiday makers home and brought predictions of a not-too-distant ice age. In 1975, however, the summer had a prolonged level of heat, said to occur for such an extended period only once every hundred years. Yet in 1976 this was repeated and a desert climate was being predicted for much of continental Europe north of the Alps. The month of June during that year was the hottest in central Europe for some 120 years. It brought the drought in Britain to near catastrophic proportions, while in the Soviet Union, where the summers are usually dry and warm, it was cold and wet.

The truth is that meteorologists are as yet little more able to explain these extremes than were their predecessors, the astrologers, in bygone centuries. In 1114, the drought made it possible to wade across the Thames, while paintings depict fairs being held on the ice of the same river. In 1473, the summer was so hot in central Germany, that fires were sparked in the forests of the Harz mountains. In 1714, only 11 inches of rainfall were recorded in London for that year (about the same as fell in southern England in December 1978 alone), while in 1719, in one region in northwestern Germany, no rain at all fell from May to October. In London, there were twenty minutes of sunshine during the entire month of December 1890, while at the other extreme 334 hours of sunshine were recorded during July 1911, when temperatures reached 100° F.

Geological, palaeobotanical, palaeozoological and glaciological research has shown that the climate has to date varied many times, moving between extremes, replacing tropical conditions with a blanket of ice. It is now claimed by some that we are moving towards another large-scale change, although there is insufficient evidence for any definite conclusions to be reached. The extreme weather conditions which we have experienced over the last few decades are not in themselves proof that a new ice age or desert climate is upon us.

Human activities are also worth considering for their effects on climatic conditions. We shall examine the effect of pollution in industrialized areas on the weather, as well as other activities which affect the climate of a particular area, in forest regions, for example, where trees have been too extensively cleared, or the land badly cultivated. In southeastern Kenya, forests were cleared for charcoal to be sold overseas. Trees increase the amount of moisture in the atmosphere by continually transpiring moisture gained from the ground. This cools the air and contributes to precipitation. In the case of southeastern Kenya, the lack of rain resulting from the removal of the trees enabled the wind to remove the topsoil which in turn meant the water table sank and a new desert has subsequently threatened human survival. Such forest clearances in Africa, South America and Asia may not necessarily have caused climatic change but they have certainly accelerated the rate of change.

While extremes or unpredictable weather provide fascinating reading, our concern is with the understanding of short and long-term forecasting of the weather in western and central Europe. Britain is under much the same weather influences as the rest of temperate Europe, as can be seen by looking at any newspaper or television weather map, or any satellite weather picture. Short-term and, in many cases, long-term weather projections are now within man's capabilities and, it is hoped, this book will provide the reader with a greater knowledge of the reasons why certain weather phenomena and patterns occur, and how to interpret official weather forecasts. At the same time, it might encourage some to try a little forecasting for themselves.

The text begins with a brief introduction to the main components of weather, where it takes place and what are generally considered to be the causes. There is then an alphabetical section, identifying the most common or spectacular weather phenomena.

This identifies the various forms of wind, cloud and storm, their causes and associated weather patterns and what subsequent weather to expect.

This is followed by a chapter on how the meteorologist goes about his business, and the instruments used in measuring elements of the weather (pressure, temperature, humidity). Having analysed these, the reader is then in a position to study graphic representations of depressions, anticyclones, high pressure systems and occluded fronts. Using a range of typical European weather maps, the reader will learn how to understand and interpret the weather systems which act on western and central Europe. The final section of the book is a compilation of the various aspects of studying the weather. Elements, conditions, and weather phenomena are all reconsidered in the light of having looked at each individually and then studied as indications on weather maps.

At the end of the main text, there is a series of short supplementary sections with information on the evolution of meteorology, the international organization of weather forecasting, cloud classification, weather charts for shipping, weather observations past and present, a list of useful addresses and, finally, a short glossary of the most important meteorological terms used in the text to help the uninitiated reader, although extensive use of cross-referencing has been made.

The earth – photographed from the spaceship Apollo 13. In the centre of the picture, visible through the cloud cover, we can distinguish the American Southwest and northwestern Mexico. The cloud formation of a depression is also very clearly depicted at the upper left of the picture. With the help of weather satellites, large cloud formations can be kept under observation continuously. Satellite photographs have become an indispensable aid to weather forecasters in, for example, the early detection of dangerous tropical revolving storms. The weather charts we see on television often become more comprehensible once we have seen the accompanying satellite weather picture.

What is weather?

Weather has many faces and many moods. It appears in many guises, each giving us the means to identify it and react to those moods. Even the most casual observer cannot but be aware of the weather's ever-changing sequence and the often sharp shift in conditions. Sunshine, cloud cover, air movement, rainfall – all these changes take place above the earth's surface, in the atmospheric shell of our planet earth.

The atmosphere is relatively dense, consisting of a number of gases the balance of which varies with altitude and latitude. Without the atmosphere, there would be no life. Animals and plants use the oxygen in the atmosphere for respiration, and plants grow by drawing on the carbon dioxide and radiant energy from the sun. This dependence of ours makes understanding the relationship between the atmosphere and weather all the more important. Also, the varying amount of water vapour present has a vital effect on the weather. It can range from a minimum of about 1 percent in midlatitudes to a maximum of about 4 percent in the tropics. Excluding water vapour, the composition of the atmosphere is:

78% nitrogen
21% oxygen
1% argon, carbon dioxide, helium, hydrogen, etc.

Nitrogen acts as an inert medium in which the oxygen can undergo its chemical changes. By the dissociation of the oxygen molecule (O_2) into atomic oxygen (O), and the combination with a further molecule to produce O_3, ozone is formed. The dissociation is the result of ultraviolet radiation; however, radiant energy with a slightly longer wavelength destroys the ozone. A continuous process of construction and destruction of ozone takes place at a height of 30km (19mi), with some sort of balance achieved. At this height, the ozone forms a protective layer, absorbing the radiant energy harmful to life.

Water vapour, which in the atmosphere forms clouds, is released from water surfaces (seas, lakes, rivers), from snow and ice-covered regions and from vegetation by evaporation and transpiration.

Dust particles, both organic and inorganic, are also present, entering the atmosphere from the earth's surface and from space. These particles have an important role in the formation of precipitation (rain, snow, etc.), acting as the nucleii on which the condensation takes place. Heavy particles of dust, soot and other air pollutants are most noticeable in the lower layers of the atmosphere. High concentrations of these air pollutants hinder the absorption of radiant energy by the earth's surface. If persistent, this can lead to a cooling and severe change in climate.

Alternatively, a persistent warming of the earth's surface could also produce climatic changes dangerous to man's existence. As the world's industrial and energy production

The sun, a fixed star, provides the earth with vital energy. Climate and weather are dependent on energy from the sun

grow, it not only increases air pollutants but it also releases large volumes of carbon dioxide. The increasing proportion of carbon dioxide in the air blocks the release of radiant energy from earth into space, with the large molecules (and water vapour molecules) intercepting the long-wave radiation. This is called the 'greenhouse effect', for, like the glass in a greenhouse, it permits the inward passage of short-wave radiation but inhibits the outward movement of long-wave radiation, raising the temperature within.

At present, it is to be hoped that these two contrasting effects arising out of human activity, one cooling the atmosphere, the other warming, are in balance – but the recorded evidence and measurements are insufficient for an assessment to be made with any certainty.

The water vapour in the atmosphere occurs mainly at lower altitudes. The heavy particles of atmospheric pollutants are also concentrated in the lower atmosphere, particularly over industrial regions, but otherwise the chemical composition of dry air shows little variation with latitude or with altitude up to 100km (62mi). Above 100km there are changes in the mixture of gases: some such as helium and hydrogen, although difficult to detect at a lower level, are present in greater proportions.

Although the composition of dry air is constant over latitude and altitude, the atmosphere is stratified according to such features as water vapour present and temperature. The formation of these layers is dependent on the circumstances of air mixing, in turn dependent upon vertical movements. In the *troposphere*, which extends up to 15km (9mi) above the planet's surface, air mixing is particularly marked. This is the zone within which weather events take place. At this level, clouds and precipitation phenomena appear. It is quite otherwise in the *stratosphere*, immediately above, where there is no vertical mixture, although there are strong horizontal flows.

Anyone who has climbed a mountain or flown in an unpressurized aircraft knows from experience that, with increasing altitude, the air becomes thinner, of lesser density and colder. Air pressure, the weight of the column of air above a given point, diminishes with height. At the earth's surface the pressure is considerable. The first measurements of air pressure were made by Torricelli, a student of Galileo. He used a glass tube, shaped in a U, to show that air pressure at sea level could support a 760mm (29.925in) column of mercury. Mercury is relatively heavy, with a density of 13.6g/cm^3. This gives in imperial measure a pressure of 14.696lb/in^2. After 1914 the British and American meteorological services abandoned imperial measures and adopted the CGS system, with the dyne the unit of pressure: 29.925in of mercury represent a force of 1,013, 231 dynes/cm^2; 1,000,000 dynes are called a bar, 1000 dynes a millibar. Barometric readings are now generally expressed in millibars.

When using a mercury barometer to measure pressure, various factors need to be taken into account. Mercury expands when heated, so that the temperature must be given at the time of measurement – or the measurement reduced to 0°C. Furthermore, as the earth is a revolving body, the centrifugal force decreases the gravitational pull to the centre. This is very marked at the equator but non-existent at the poles. As a result, the height of the mercury column varies at different points on the earth's surface.

760mm (29.925in) mercury column = 1013.2mb

An air pressure of 1000 millibars corresponds to a mercury column of 750mm (29.528in) at sea level at latitude 45° with a temperature of 0°C.

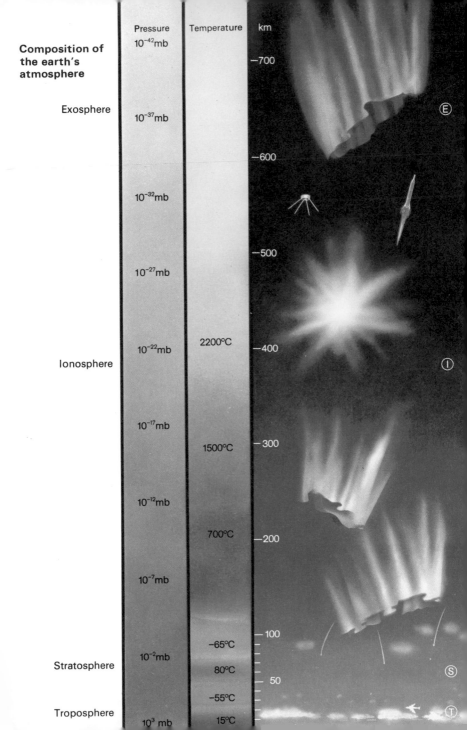

Composition of the earth's atmosphere

Exosphere

Ionosphere

Stratosphere

Troposphere

Pressure	Temperature	km
10^{-42}mb		−700
10^{-37}mb		
		−600
10^{-32}mb		
		−500
10^{-27}mb		
10^{-22}mb	2200°C	−400
10^{-17}mb	1500°C	−300
10^{-12}mb	700°C	−200
10^{-7}mb		
		−100
10^{-2}mb	−65°C	
	80°C	−50
	−55°C	
10^3 mb	15°C	

The column of air above the earth's surface decreases with a gain in altitude and the air pressure correspondingly falls with altitude. The actual density of the air also decreases, as pressure is reduced. On average, with increasing height, the pressure falls as below:

height (m)	pressure (mb)	temperature (°C)	density (g/m³)
0	1013	15.0	1225
111	1000	14.3	1212
988	900	8.6	1113
1949	800	2.3	1012
3012	700	−4.6	908
4206	600	−12.3	802
5574	500	−21.2	692
16180	100	−56.5	161

The relationship between altitude and the height of the barometer provides the theoretical basis for the altimeter and, indeed, altimeters are used in pressure measurement. The barometer/height formula, however, is not simple. The lower atmosphere is denser and at 5.5km (3.42mi) altitude pressure is only half that at sea level. Furthermore, there is a cooling in the lower atmosphere as altitude increases, so that temperatures far below freezing point are reached. This pattern is not, however, maintained consistently with increasing altitude. In the stratosphere, a slow upward incline of temperature begins and then it drops again before rising to many times the surface temperature. The irregularities of density distribution in the upper atmosphere are influenced by temperature, and the variation in temperature is in turn due to differences in the absorption of the sun's radiant energy.

air temperature at 10km (6.21mi) altitude − 60°C, at 80km (49.68mi) − 90°C, at 50km (31.05mi) altitude 0°C, at 150km (93.15mi) altitude + 1000°C

Information gained with the use of rockets and satellites shows that the upper atmosphere is layered, as seen on page 15, and this applies particularly to the *ionosphere*, between 60km (37mi) and 2000km (1242mi) altitude. Radiant energy from the sun passing through the ionosphere displaces electrons from the gases present, producing charged particles or ions. To do this, the wavelength of the radiant energy must be lower than a critical value which varies from gas to gas. In the E and F_1 zones on the diagram the radiant energy is probably x-rays, and the gas from which ions are produced is probably oxygen. The entry into the ionosphere of charged particles, probably hydrogen ions from the sun, is associated with the polar lights. The ionized layers reflect short-wave radio waves, and make possible radio transmission over thousands of kilometres. To an increasing extent, this role of the ionized layers is being superseded by communication satellites, which are less affected by atmospheric disturbances.

Transmission and penetration of waves in the earth's atmosphere

1 X-rays 2 ultra-violet radiation 3 visible light (optical window) 4 infra-red radiation 5 radio wave radiation from space (radio window) 6 ultra short-wave radio waves 7 short-wave radio waves 8 middle-wave radio waves 9 long-wave radio waves

Although weather phenomena are concentrated in the lower levels of the atmosphere up to 15km (9.3mi) in height, we need to look, as well, at the upper levels. Radiation from space, mainly but not exclusively from the sun, passes through the upper atmosphere and, in doing so, is altered. The best known example, which has already been mentioned, is the absorption of ultraviolet radiation with a wavelength of less than 300 nanometres (0.003m or 0.118in). The x-rays, probably responsible for ionization of oxygen in the E and F_1 layers, would be absorbed in the process. The thin atmosphere has an effect comparable with a lead sheet metres thick in its absorption of radiant energy. However, some of the sun's radiation reaches the lower atmosphere, where it provides the energy responsible for the weather.

Only a part of the solar radiation admitted into the atmosphere reaches the earth's surface. In its passage through the atmosphere, the radiation has little heating effect – but when it reaches the earth's surface it is absorbed, warming the earth which, in turn, heats the atmosphere from below.

The various parts of the earth's surface receive differing amounts of radiant energy, depending on latitude, time of day and season of year. The latitudes near the equator receive much more solar radiation than do the latitudes near the pole. In summer the incoming radiation is greater than in winter, and of course summer in the northern hemisphere corresponds with winter in the south, and vice versa.

The effect of incoming radiation also depends on the nature of the surface. Land and sea differ greatly in their absorptive capacities. Land warms quickly and releases, in turn, a great deal of heat to the atmosphere. It also can cool down considerably, a characteristic it shares with the lower atmosphere. The process is most effective when the earth is cool and dry. Soil covered with forest, warms and cools more slowly. The many types of surface forms mean that, on occasion, notable temperature differences can be registered in a very small area. Where ice and snow cover the land, the high reflection rate and the latent heat of melting reduce the amount of energy available to warm the ground and air. The temperature of a snow sheet therefore drops sharply on a clear night, more sharply than the temperature on uncovered ground. During the day, the thermometer does not rise above zero.

The oceans also have an important function as a reservoir of heat. The cycle of heat gained and lost by water begins when the sun's radiation penetrates much deeper into water than into the land surface and the heated mass is therefore much greater. The rise in temperature is slower than that of the land and the temperatures reached are not so high but, at night and in winter, the water does not cool so quickly as the land. As the surface water cools and sinks, it is replaced by warm water from below which the sun's radiation had penetrated during the day. Considerable heat is taken up as water vapour, i.e. by evaporation, but without immediately warming the air. Only when clouds are formed is this heat released by the water vapour. The effect of large lakes on

---

Radiation in the earth's atmosphere

Scattering, absorption and reflection weaken the solar radiation even in the ionosphere and stratosphere. Of the radiation reaching the tropopause, 34% is reflected and scattered. A further 19% is absorbed by water vapour and carbon dioxide, and only 47% reaches the surface of the earth. Of the long-wave radiation from earth,

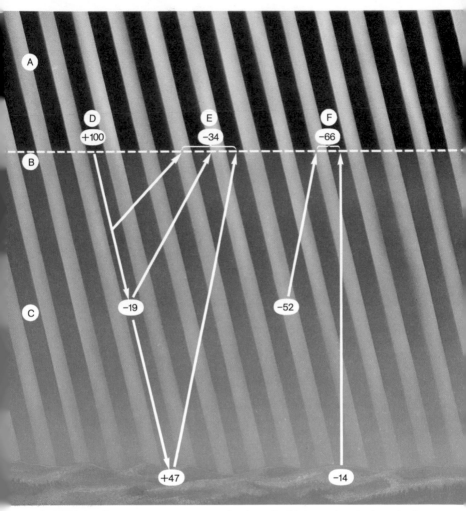

only 14% is lost because water vapour and carbon dioxide have absorbed much of the radiation and returned it to the earth once more – the 'greenhouse effect' already mentioned. In total, the earth's surface has a surplus of 33% and this energy is returned to the atmosphere. The atmosphere receives warmth from below and this warmth is radiated above and below (52% of the long-wave outradiation). A = stratosphere, B = tropopause, C = troposphere, D = total incoming radiation, E = loss resulting from absorption, reflection and scattering, F = outradiation from the earth and from the troposphere.

Particularly at dawn and at sunset the sky is magnificently coloured. This is caused by the scattering and reflection of the sun's radiation in the atmosphere

the weather is similar to that of the sea. When winds are onshore, the oceans moderate the climates of adjacent land masses, as with the oceanic climate of Western Europe, but their effect is lost when easterly air movement brings the bitter cold of the continental interior.

The Greeks of the classical period had already concluded that the sun's radiation was responsible for wind movement but the explanation of the process had to await the advent of modern physics.

Only part of the sun's radiation is absorbed by the earth's surface. Some of it is reflected immediately. Even before the earth is reached, the clouds form a highly reflective layer. From mountain tops or from aircraft, it is sometimes possible to see the upper side of a closed cloud sheet shining a brilliant white because of the high degree of reflection.

Light of various wavelengths has very different visible characteristics when seen in the processes of absorption, scattering and reflection. The sun's radiation has a spectrum that extends from ultraviolet to infrared. The strength of the radiation in various parts of the spectrum gives a reference point for the temperature of the sun. The greatest intensity of radiation occurs at a wavelength of $4700°$A (angstroms are the units of measurement for the length of lightwaves; 1 angstrom = $1/100,000,000$th of cm). The surface temperature of the sun is just below $6000°$K (Kelvin is the scale used in measurement of absolute temperature).

In addition to the radiation that warms the earth's surface, the sun also produces short-wave ultraviolet and x-rays, and an atomic particle radiation of electrons and protons similar to that received from outer space. The function of various atmospheric layers in reducing the degree of penetration of these harmful radiations has already been mentioned.

As it penetrates the troposphere, the incoming radiation is absorbed, reflected and scattered. Cloud formations and the carbon dioxide content of the air influence absorption and reflection. The scattering is a result of a change in direction of the incoming radiation on impact with air molecules. The blue end of the spectrum (short-wave section) is more strongly scattered than the red (long-wave section).

The scattering of sunlight by the molecules of nitrogen and oxygen in the atmosphere is a selective effect most marked at the blue end of the spectrum. As a result, the sunlight is not completely white but appears yellow to the observer on earth.

Absorption, reflection and scattering together cause the loss of over half the incoming radiation. Only 47% of this radiation reaches the earth's surface, and only a part of this is absorbed. Energy is immediately lost from the surface as long-wave radiation. This outward radiation is effectively reduced by further absorption and reflection in the atmosphere, the result of the presence of carbon dioxide and water vapour. After all these losses have been accounted for, about a third of the incoming energy remains on the earth's surface to initiate the weather events.

The energy of this absorbed radiation is released to the atmosphere by heating. Were this not so, temperatures of the surface would rise to levels harmful to life.

The heating of the atmosphere is effected in two ways. First, temperature becomes more evenly distributed by the movements upwards and downwards of air masses. Second, there is cooling by evaporation. More than 75% of the earth's surface is water, and the removal of heat from the earth's surface into the atmosphere by evaporation is extremely important.

Clouds forming a reflective layer for the sun's radiant energy

The position of sun at various seasons

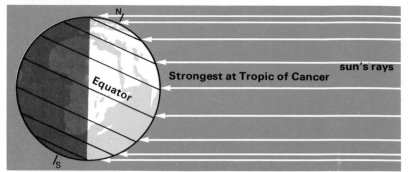

Position of earth on 22 June, beginning of summer in northern hemisphere. The average duration of sunshine at 47°N is 15 hours 50 mins., at 53°N 16 hours 52 mins., at the north pole 24 hours. Winter begins in the southern hemisphere.

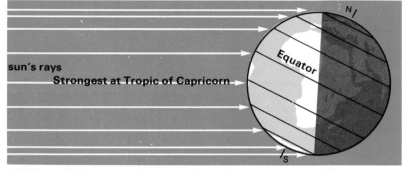

Position of earth on 22 December. Winter begins in the northern hemisphere. The average duration of sunshine at 47° is 8 hours 26 mins., at 53°N 7 hours 29 mins., at the north pole 0 hours. Summer begins in the southern hemisphere.

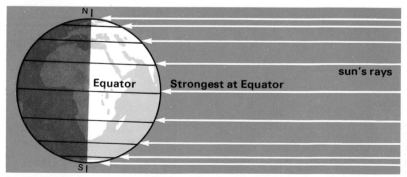

Position of the earth on 21 March and 23 September. Beginning of spring and of autumn in the northern hemisphere. The average duration of sunshine is 12 hours at all latitudes. In the southern hemisphere, autumn begins on 21 March, spring on 23 September.   N = north pole   S = south pole

The atmosphere itself cools very quickly. Without the continuous movement of heat from the earth's surface, the atmosphere would progressively cool.

Meteorological events are described in terms of the weather *elements*, observed and measured at weather stations. The meteorological elements are air pressure, air temperature, wind, humidity and density.

These elements are discussed individually in the section on basic meteorology, p. 95. Their measurement, together with the mathematical development of physical models, for the explanation of processes in the atmosphere and for short medium and long-term forecasts, is a complicated process. In recent years, weather satellites and computers have made both observation and evaluation easier while, at the same time, significantly improving prediction.

Meteorologists must observe the condition of the atmosphere at a selected site, not for days or weeks, but over very long periods in order to have a suitable basis for forecasting. The daily, monthly and annual means of air pressure, etc., are required, in order to establish the average sequences of weather associated with a district. It is then possible to talk of 'climate'. Climatic elements are more wide ranging, including, in addition to meteorological data, information such as duration of sunshine, location (using geographical co-ordinates), and vegetation.

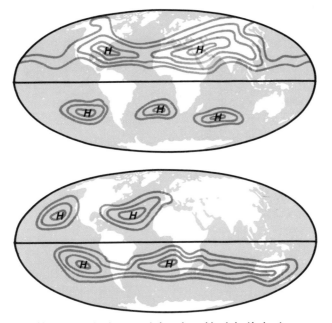

Air pressure in January (*above*) and in July (*below*)

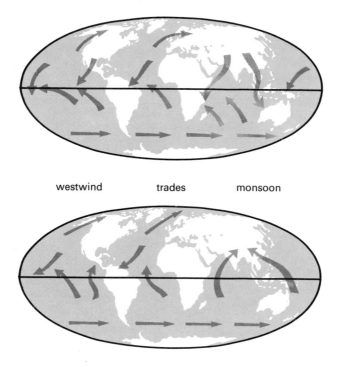

westwind   trades   monsoon

The most important wind directions in January (*above*) and in July (*below*)

Even without considering the influence of location on the climate, geographical site is of prime importance. It determines the relationship of inward and outward radiation, that is to say, the radiation 'budget'. The influence on climate and on climatic change arising out of human activities has already been mentioned. Observations show clearly that buildings, reservoirs and forest clearances do influence regional climate, but it is still not possible to project categorically the changes caused by man's assault on nature. Observed climatic changes could result from changes in the positioning of the earth's axis which, in turn, could influence man's use of fuels, leading to changes in the pollutants being given off. Perhaps change can best be shown with one example – with every 1°C fall in average temperature, the summer in the middle latitudes – the period when crops can ripen – is shortened by a week.

Two photographs on the same day, an hour apart, of the Great St Bernard's Pass. The upper picture shows the northern side of the Alps (Wallis), the lower picture, the southern side of the Alps (Aosta valley). Distinctive weather can often be experienced on either side of a medial ridge. Mountains, and in particular the Alps, form an impediment to moving air. Mountains act as a dam in keeping back or redirecting air masses, and consequently affect the weather.

Cloud forms are extremely variable. Föhn clouds over the Bavarian Alps (*above*). Towering cumulus cloud over the Hardangerfjord in Norway (*below*).

Weather: what we see

A practical guide to identifying the natural phenomena

This section is devoted to the natural phenomena which we can identify and which are significant in determining weather patterns and climatic conditions. Each feature, whether it be something as seemingly everyday as blue skies or moonlight, or something as dramatic as lightning or a highly-coloured sunset, is treated in the same manner. The phenomenon will be described under four separate headings: **Recognition, Causes, Effects on weather**, and **Forecast**.

Under **Recognition** and **Causes**, the reader will find descriptive details of the physical characteristics and then a look at the causes of the phenomenon. These will assist in its identification. **Effects on weather** places the phenomenon in context, relating it to the weather sequence, with an explanation of its influence on weather and climate.

Under **Forecast**, there is an assessment of the likely nature of subsequent weather indicated by the presence of the particular feature. Remember, however, that weather is dependent on many factors and one phenomenon will simply point to a tendency.

Observing the weather is something everyone can do, in town or country alike. Photographers will find cloud formations and highly-coloured atmospheric phenomena particularly attractive subjects. You may want to back up simple observation by taking your own measurements of the most important meteorological elements such as pressure and temperature. Simple and cheap equipment is available as shown on p. 110. Information about regular weather recording is given in the section on weather observation. The keener reader who wishes to test the value of a long term series of observations should get in touch with a local weather station (see list of U.K. weather stations, p. 221).

Air streams - high altitude

Recognition. In a clear blue sky, the watchful observer will occasionally see thin wisps of cirrus cloud (see p. 34). Sometimes these wisps appear to hang motionless with little change in shape. Often, however, they are torn by the winds, developing a drawn-out appearance with a hooked end. It is easy to see that this cloud movement is the result of an air stream moving at the level of the cloud. Similar observations can be made of cirrostratus cloud (see p. 86), usually indicating the approach of a warm front. The arrival of cirrostratus gives a milky appearance to the heavens. Occasionally, these sheet-form clouds are drawn out into bands, these in part being whirled. Here again, as with the wisps of cirrus cloud, there is evidence that a strong air stream is driving the clouds.

Causes. Weather events are not limited to the air masses near the ground. Even at an altitude of 5 to 10km (3 to 6mi), warm and cold air masses are in contact, and their temperatures and pressure differences are emphasized. As a result, very strong air movements take place at astonishing speeds. At an altitude of 5km (3mi), speeds of 300 to 400km/hr (185 to 250mph) have been measured, and such speeds can considerably affect aircraft flying at those heights. The 'jet stream' is so strong that it has to be allowed for in the planning of air routes. It can, for instance, dramatically affect the duration of trans-Atlantic flights. Flying with a jet stream as a tail wind can cut an hour from the crossing with subsonic planes. The movement of air at high altitudes sets off a circulation of air from ground level to the higher levels of the troposphere. From depressions, air reaches the higher levels, later to be returned towards the ground in high pressure systems.

Effects on weather. In Europe the main route for high-altitude air streams lies over Ireland, England, northern Germany and Poland. It has been shown, however, that over long stretches this streaming coincides with depressions, and that sudden changes in direction of the air streams can occur together with changes in the direction of movement of depressions. Eddies in the jet stream are of special significance to weather, for they give rise to particularly intense depressions, accompanied by strong winds and heavy rain.

Forecast. Movement and alteration in the shape of cirrus and high-level stratus clouds give some indication of air streaming at high altitudes. In particular, these high clouds mark the approach of a depression. If, for example, the cirrus drifts from west to east, the lower clouds may still be moving in another direction. A change to poorer weather is indicated only when the movement of the cirrus cloud is at an angle to that of the ground winds as indicated by the movement of the lower cumulus and stratus cloud. If the trajectory of the cirrus crosses that of the lower stratus and cirrostratus from left to right (when the observer has the ground wind to his back), then a deterioration can be anticipated, since the lower winds are backing. If the trajectory of the cirrus crosses that of the ground winds from right to left, then a depression is departing. The third possibility, the coincidence of high and low level winds, or their opposition, indicates no change in the weather.

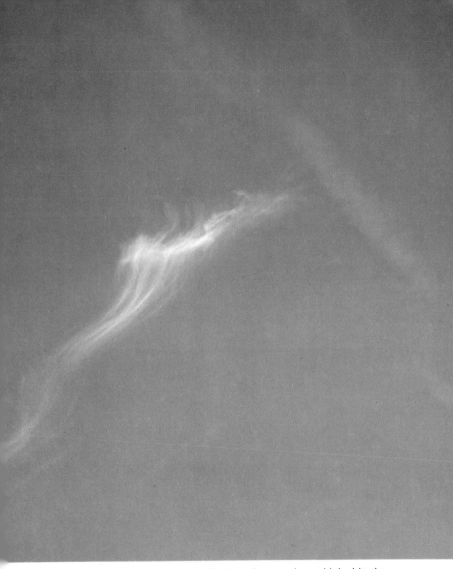

The movement of soft cirrus clouds, indicating air streaming at high altitudes

Blue skies

Recognition. This is a particularly impressive phenomenon in mountainous country. On a calm and cloudless day the blue of the sky seems intense and absolutely pure, in winter usually even more so than in summer. From the valley, although the good weather and sunshine remain, the sky appears less blue. In towns, even on very fine days, there is an impression of a paler blue, more white or grey, than the intense blue of the mountains. In different parts of the world, there are marked variations in the intensity of the blue – it is particularly striking, for example, in the tropics.

Causes. The blue colour of the sky is the result of the interplay of sunlight and the atmosphere. Without the atmosphere, the skies would be dark, with sun and stars seen at the same time. The sun's radiation is scattered widely by the air, by the dust and by water droplets, which diffuse its light so that it swamps out the light from the stars. Why is the sky blue? When we see all the colours in the spectrum, we see white. In the atmosphere, the short-wave blue light particularly is scattered and becomes more noticeable than the reds or greens.

There are also similar simple but important grounds for the variation in intensity of the blues. The opaqueness of the atmosphere is caused by dust and water droplets. The more dust and water droplets present, the paler the blue. The heavens appear then, even in dry sunny weather, more white and grey than blue. Conversely, the absence of dust and water droplets gives a deeper blue, as in the mountains.

Effects on weather. Blue skies are not always associated with high pressure. They can occur in the lower pressure area between two highs as in the föhn weather (see p. 46). In these conditions, visibility is often excellent. Blue skies often occur before the arrival of a warm front (see p. 90). After a troubled period, a sudden break in the cloud cover can reveal a misleading, almost unreal, blue sky. The weather is unstable and low pressure is predominant. Conversely, with a high pressure in summer, the skies can be pale, indicative of an increasing dust content. Visibility is poor and distant views are hazy. There is here a seasonal difference in that blue skies are generally more intense in highs occurring in spring and autumn than those occurring in summer. Although a high pressure may predominate, blue skies in winter are normally less deeply coloured because of the dust associated with inversions (see p. 62). Only at high-altitude observation stations is the blue more intense.

Forecast. The depth of blue coloration in the sky can be indicative of forthcoming weather. For lowlands, there are three firm rules:

1. Extremely deep blue, together with oversharp visibility, shows a changeable weather condition. A sudden deterioration with a windstorm and rain cannot be discounted.
2. Medium to light blue means a continuing spell of good weather.
3. Increasing change from blue to white or grey, accompanied by haze, is indicative of change in the form of a depression.

Blue skies with isolated wisps of cirrus clouds

Breezes – land and sea

Recognition. At the seaside during sunny weather, one often notices that in the early afternoon a pleasant cool breeze begins to blow in from the sea. The cool sea air pushes under the warm land air, lifts it and creates small cumulus clouds (fair weather cumulus) which disappear almost as quickly as they form. Similar observations can be made in mountainous inland regions. On warm days, about midday or in the early afternoon, an up-slope wind develops, blowing from the valley. Small cumulus clouds are formed in the same way as on the coast and later disappear.

On the coast, the sea breezes die away gradually in the evening, and later, at night, a land breeze springs up, blowing towards the sea. Anyone out before dawn will find this land breeze blowing quite strongly. Similarly, in the mountains, the up-slope breeze dies away in the afternoon, and is replaced in the early evening by a cold down-slope breeze. A further air movement develops on the valley floor and, in the second half of the night, a stiff breeze blows down the valley.

Causes. A principle of great significance to weather is that land and sea warm up and cool down at unequal rates. As the sun becomes higher, the land warms more rapidly than the sea. The warming of the lower layers of the atmosphere over the land leads to a fall in pressure and the establishment of a pressure gradient from sea to land. This makes the sea breeze begin to blow.

In the mountains the valleys and mountain slopes warm up and cool down at different rates. During the day, the valley air is heated to a greater intensity than that on the mountain side and, being heated but constricted, expands and rises up the slope.

At night the land cools more rapidly than the sea. A lower pressure is established over the sea and the land breeze blows towards it. During the night, cold dense air from the mountains flows down the slope into the valleys. A pressure gradient is also formed near the exit of the valley, which is lower and warmer. This pressure gradient is responsible for the second, down-valley wind. This wind activity in good weather, both by the coast and inland, calms down only when there are no temperature differences between land and sea, or mountain and valley – a condition which can be expected in the hours immediately after dawn and just after sunset.

Effects on weather. Land and sea breezes, as well as valley and mountain winds, are local phenomena. They occur only on clear summer days when the weather is not dominated by larger disturbances, and are most highly developed when the variation in temperature between day and night is greatest. The mechanism is more effective in lower than in higher latitudes, and in high mountains than on plateaus. Local conditions are very important. In addition to these daily systems, there are seasonal winds resulting from the same basic causes. The most obvious example is the Indian monsoon, which develops in response to the contrast between hot land and cooler sea in summer, cool land and warmer sea in winter. There are also monsoon-like effects in Europe. Northern weather systems in spring and early summer are caused by the warming of the continent, and bring cool moist sea air from the northwest. Southern weather systems in late summer and autumn are caused by the cooling continent, bringing dry air from the southeast.

Forecast. Daily land and sea breezes are good weather winds. This is confirmed by the short-lived cumulus clouds.

A warm up-slope mountain wind is marked by the presence of cumulus clouds in the northern Alps.

Cirrus clouds

Recognition. Cirrus clouds float in blue skies looking as though they have been whisked there by the wind and scarcely warranting observation. This would be a mistake. In sequence with cirrocumulus (see p. 38), and with cirrostratus (see p. 86), cirrus clouds provide evidence that a change in the weather is taking place. Indeed, with the arrival of cirrostratus clouds, the skies become overcast and a change is imminent. Strong winds at great heights are shown by the presence of cirrus clouds when they are drawn out in long plumes hooked at the end. In some parts of Europe, these forms are described as 'hooked cirrus'. If the plumes are drawn out to greater lengths, they give the impression of being driven and pushed about by the wind.

Causes. Cirrus clouds are formed at heights of about 6000 metres (19,600ft), and belong to the high clouds types with cirrocumulus and cirrostratus. When cirrus clouds appear without cirrocumulus and cirrostratus, they have a very delicately arranged structure. In form, they resemble 'ice banners' and 'snow banners', by which names they are sometimes known, for, as the highest clouds in the troposphere, they are pure ice clouds, consisting solely of ice crystals. The movement and form of cirrus clouds is interesting, providing a guide to air movements and wind strength in the atmosphere. They indicate the direction of the jet streams (see p. 28), of significance to weather events over large regions. The temperature of cirrus clouds is below $-40°C$.

Effects on weather. Cirrus clouds are the first indicators of certain weather events. Most significant is the association of cirrus with an advancing depression, before the warm front. The greater the air movement, the more disturbed are the cirrus clouds. This warning of an advancing depression is most reliable at sea or along the coast. In inland areas, the process is slower and the cirrus may be present for several days before the depression arrives. In contrast, cirrus clouds, already festooned with low-hanging cumulus approaching at considerable speed, can also herald an inbreak of cold air.

Forecast. The thickening of cirrus cloud to a cirrostratus sheet indicates a reasonably rapid approach of a warm front with rain. The same applies to the sudden appearance of cirrus together with aligned cumulus cloud as harbingers of cold air from the north or northwest. This is usually accompanied by heavy rain and, eventually, frontal storms. Cirrus clouds are also present in good weather, drifting slowly from east to west and dissolving during the day. Other cirrus clouds seem almost still but this is the result of their great height, making movement difficult to detect from the ground.

Clearly-shaped cirrus clouds. Their form is extremely variable, ranging from feathery wisps to a veil covering the sky.

Coronae

Recognition. Around the sun or moon we can sometimes see a ring of white or yellow, occasionally bounded on the exterior side by a brown to violet rim. This cannot be confused with halos (see p. 52), for a corona consists of a coloured disc with the sun or moon in the centre and apparently touching, and not an encircling ring. Furthermore, the corona is of much smaller diameter than the commonly observed small ring. Coronae, particularly those around the sun, are often coloured, although not so brightly as a rainbow (see p. 78) with red on the outside edge, and are always round, with a concentric structure. As distinct from halos around the moon, lunar coronae are clearly coloured at full moon. When observing a corona around the sun, it is advisable not to look directly at the sun, but to observe it through dark glasses or, if none are available, to obscure the sun with the corner of a building. Coronae are also observed around Venus and the brighter fixed stars, such as Sirius.

Causes. Coronae differ from halos not only in appearance but also in the associated physical conditions. The halo has its origin in the refraction of sunlight by ice crystals in middle and upper parts of the troposphere, whereas coronae are caused by refraction by water droplets in middle-altitude stratus clouds (see p. 86). The droplets hinder the passage of the widening beam of radiant energy from sun, moon or other heavenly body. The light is described as scattered.

Effects on weather. Since coronae are the result of scattering of light by water droplets, very high clouds, consisting of ice crystals, are excluded – that is to say cirrus, cirrostratus and cirrocumulus. Coronae are more likely to be formed in association with altostratus clouds, consisting of ice crystals and water droplets, or in association with altocumulus, very level cumulus clouds consisting predominantly of water droplets. Vertical movement can develop in altocumulus, but often such clouds are fibrous and diffuse, resembling stratus clouds. In them, scattering is particularly marked, making them iridescent (showing shifting rainbow-like colours). The occurrence of altostratus and of altocumulus is normally linked with weather influenced by a depression.

Forecast. The forecast based on coronae is similar to that based on halos, around the sun or moon, since both types of phenomena indicate moisture in the atmosphere. The significant difference is that the change in weather indicated will occur more rapidly with a corona, since the moisture in the atmosphere is already at a lower level. Generally, the rain will come shortly after the warm front. In winter, however, cold weather may persist for some time after a corona and the change, when it comes, may be accompanied by heavy snowfall.

High stratus clouds produce a corona around the moon

Cumulus clouds

Recognition. Cumulus clouds can be seen developing when a storm builds up. At first, there are gently-rounded woolly masses, but then suddenly great towers rise rapidly upwards. These are clouds with sharply-defined edges; but the towers formed when the weather is stormy have ill-defined edges, appearing to smoke. Not all cumulus cloud leads to storms. In summer, fair weather cumulus come and go quickly, assuming rounded forms and not characterized by the tower growth. The structure shown opposite is typical, the domes or towers often blinding white in the sunlight. The bases are frequently grey, often very dark, and normally horizontal. The bases in fact indicate the condensation level. Cumulus clouds change shape very quickly, on average in a matter of five to twenty minutes. They are grouped according to height as follows:

1. High cumulus: Cirrocumulus (Cc), at heights of over 6000 metres (19,685ft). Usually well-ordered in ranks of small white cloud balls, with additional tiny white patches of cloud. Ice clouds.
2. Medium-height cumulus: Altocumulus (Ac), at heights between 2000–6000 metres (6560–19,685ft). Appearing like small sheep, or scales, also in rolls and ball-like. Fibrous, occasionally diffuse.
3. Cumulonimbus (Cb), at heights between 1000–5000 metres (3280–16,400ft). The large storm towers; mixed clouds consisting of ice and liquid water. Known for their sleet and hail showers, and squalls.
4. Low cumulus: Stratocumulus (Sc) at heights below 2000 metres (6560ft). Low-hanging rolls of cloud: colour grey or pale grey with dark divisions.

Causes. Cumulus clouds, in contrast to stratus clouds, are usually found at considerable heights. They originate in the rapid ascent of warm air, the inbreak of cold air into a warm-air mass triggering off strong vertical movements, often producing towering cumulus – a frequent feature of April weather. They are also known as thermal clouds. The rising wind or 'thermal', as measured in storms, rises as high as 10km (6·2mi).

Effects on weather. Cumulus clouds can occur in association with various conditions. They are formed most frequently because of the rapid ascent of warm air, as, for example, the short-lived cumulus formed in the morning during periods of good weather. The same rapid ascent also applies to the cumulus clouds that mark a coming storm, and with the cumulus cloud formed in northwestern Europe resulting from inbreaks of cold air. The heavy cold air pushes below the lighter warm air and pushes it upwards quickly, forming large cumulus clouds.

Forecast. The smaller clouds, formed for a short period in the morning, are indicative of continuing good weather, but the towering cumulus with rough edges at the top, formed after an extremely rapid ascent, indicate the coming of a storm.

Cumulus clouds, here fair weather cumulus

Dawns and sunsets (highly-coloured)

Recognition. Shortly before sun-up and after sundown, in the dawn and evening twilights, highly-coloured skies often appear, the best known being red dawns and red sunsets. In clear weather, the red predominate, changing to yellow and grey according to the amount of dust present. The reddening of the clouds may be exceptionally intense, and the watchful observer can sometimes see colouring on the opposite horizon, i.e. in the west at dawn and in the east at sunset. In mountains, slopes facing the west take on their own reddish tones and sunset glow. The tint actually veers more towards yellow or towards red according to the condition of the atmosphere.

Causes. Twilight is the result of reflection and scattering of the sun's rays in the upper atmosphere after or before the sun has set or risen respectively.

type of twilight	sun below horizon	duration at $+50°$ N
civil	6°	30–40 min
nautical	12°	60–80 min
astronomical	18°	about 120 min

At the end of civil twilight it is still possible, given cloudless conditions, to read, whereas at the end of astronomical twilight no trace of the sun's rays is observable in the atmosphere. Because of the steep path followed by the sun, twilight in the tropics is very short, and conversely very long in the polar zone. North of 49°N, the zone of midnight twilight begins, in high summer the sun not reaching more than 18° below the horizon. The first purple light occurs when the sun is about 3° to 4° below the horizon, with a cloud-free sky. This is 15 to 30 minutes before dawn or after sunset, with the counter twilight on the opposite horizon. When the sun is 8° to 10° below the horizon, the second purple light occurs. The purple light is responsible for the colouring of the clouds, and of rock walls on mountain crags. As the sun falls below the horizon, it casts a shadow of the earth on the eastern sky (and a similar shadow in the west before dawn), forming the 'twilight arch' between the advancing darkness and the still-lit sky.

Effects on weather. The twilight colours are pure and true in a cloudless sky. Changes in colour, particularly towards red-yellow or yellow-grey, indicate increasing amounts of dust and moisture in the atmosphere. Clouds help to diversify the patterns of colour. Low-hanging clouds at sunset show marked contrasts between their east-facing sides, black or grey, and the west-facing sides which often show a mixture of reds, yellows and white. Similar contrasts and colorations can occur when there is a rapid rise in temperature, for example when a front goes through (see p. 44) or after a storm.

Forecast. Every discoloration of sunset skies with yellow or white shows the beginning of a disturbance and, if clouds are present in the lower layers, it indicates a deterioration. The old saying, 'Red sky at night, shepherd's delight, Red in the morning, shepherd's warning', is not a reliable guide. A red sunset usually indicates good weather only if it occurs in an undisturbed and cloud-free sky.

Coloured sunset in altocumulus (*above*)

Twilight colours with thick cirrus (*below*)

Depressions

Recognition. Isolated cirrus clouds (see p. 34), followed by cirrostratus and altostratus clouds, indicate a change in the weather. By the following day the clouds are thicker, a grey sheet of cloud is covering the sky and drops of rain or snow are beginning to fall. This disturbance is accompanied by a fall in pressure and a gradually freshening wind, first southeasterly but veering to south and southwest. These events can extend over one or two days, or alternatively can quickly give place to stormy, showery rainfall. Rolls of cumulus clouds appear to the west, the southwesterlies become squally and veer further to northwest. Within a few hours, the cumulus clouds become larger and more threatening and, shortly afterwards, rain showers occur, often with sleet and hail and accompanied by thunder and lightning. Clouds, rain and wind dominate the area.

Causes. European weather is the result of warm eastward-moving air streams from the subtropics, and cold westward-moving air masses from the Arctic. The boundary between these two types of air mass is the polar front. This is in no sense a straight line. Because of the differences in density and temperature, the cold air often penetrates the warm air at intervals. The waves of moist subtropical air and the dry cold arctic air interlock with each other and form low pressure areas (cyclones). It takes time for these two air masses of completely different nature to mix and, at the boundary between them, vortices develop, with a characteristic weather sequence warm front and cold front, see chart, p. 189).

Cyclones have initially an extent of a few hundred kilometres/miles. In their lifetime of two to four days, they increase in size to 1000–3000km (600–1800mi). The depressions in Europe generally move from west to east. The jet stream (see p. 28) is important in their development.

Effects on weather. Unless disturbed, the following sequence applies:

1. Cold air inbreak, easterly
2. Warm oceanic air (warm air sector) is pushed by the southwesterly winds on to the cold air lying to the east (warm front)
3. Second cold air inbreak, westerly
4. Westerly cold air pushes into the warm sector (cold front)
5. Warm air sector becomes smaller as the western cold air moves more rapidly eastwards
6. Cold front and warm front meet and the different air masses are totally sucked into the vortex (occlusion), bringing an end to the cyclone.

In reality, the events do not follow this exact pattern or sequence. Mountains for example, speed up the meeting of the cold and warm fronts and the end of the cyclone. The distribution of land and sea affects the existence of the cyclone and the routes taken by the jet streams, thus influencing the trajectories of cyclones. Indeed, cyclones follow typical trajectories across given areas.

Forecast. The air in the depression lifts and, with increasing height, expands and cools. The condensation of water vapour leads to clouds and precipitation.

Thick altostratus and altocumulus cloud — harbingers of a weather change that can bring rain

Depressions – following weather patterns

Recognition. The cold front in a depression (see p. 42), which can be recognized by the massive line of cumulus cloud, brings rain and wind – the so-called frontal weather. At the observer's position, there are stormy west to northwest winds, a fall in temperature and rain showers. As the front passes, the sky clears and visibility improves strikingly; the pressure rises and the wind, while remaining fresh, veers. Then comes the weather associated with the rear of a depression, blue skies and sunshine but not usually of any long duration. Rain soon recurs and the weather remains changeable. Usually a very rainy period of poor weather ensues with heavy showers and stormy winds, reaching gale or even hurricane force at sea.

Causes. The wind system of a depression, which forms the vortex, consists of southerlies on the east and northerlies at the rear of the depression, the west side. There are warm and cold air masses with the associated typical weather sequences. As the cold front passes, subsiding air prevails resulting in a break in the clouds. A short-lived period of better weather follows, but the sun soon warms the lower layers of cold air, which are already rich in water vapour because of the preceding rain. Vertical air movements begin, leading to the formation of cumulus cloud, and the rain begins again. The passage of a depression does not always lead to any rise in air pressure, but rather it may lower it. With higher windspeeds more cold air is brought in, storm winds develop and very heavy showers follow in a rear trough.

Effects on weather. The weather at the rear of a depression is part of a well-recognized sequence. This is particularly true for westerly weather conditions and northerly weather conditions. The break-up of the clouds and the arrival of better weather comes more quickly and with less disturbance in Britain, France and north-central Europe than in the central and Alpine regions. The Alps form a great barrier to moving air masses. The northwesterly or northerly winds behind the cold front drift the clouds together, thus forcing the clouds to rise and cool, and resulting in persistent rain, lasting a day or more. Instead of a simple cold front, a sequence of cold fronts in military-like formation can occur, the better weather coming after the passage of the last of the sequence.

Forecast. There is a general rule that ensuing weather will be more unsettled if the clouds break quickly. A longer period of calm can be due to a ridge of high pressure (see Occluded fronts, p. 192), but this is usually only for one or two days. A sequence of depressions must obviously bring another deterioration in the weather. Only rising air pressure and a veering of the wind to the north or northeast indicates a change.

Altocumulus clouds during passage of a rain belt (*above*)

Massive cumulus clouds during passage of a shower belt (*below*)

Föhn (fall) winds

Recognition. These winds, originating in the Alps, occur suddenly and in strength in the valleys. They have given their name to similar winds elsewhere. They are dry and associated with steep rises in temperature of 10–20°C, extremely clear visibility and blue skies with lens-shaped clouds. They are particularly marked in winter, when their warmth may melt the snow and make it seem that spring has arrived overnight.

Similar winds can be observed elsewhere: on the south side of the Alps, in Greenland and Spitzbergen, east of the Rockies in North America and of the Andes in South America, as well as in Japan. Even in Britain the effects of Föhn winds are occasionally experienced. They occur in eastern Scotland but, since the relief is less dramatic than in the Alps, the effect is less noticeable.

Causes. Föhn or fall winds take place in mountain ranges. The range forces air streams to rise to the mountain peaks, and then descend into the valleys beyond. With elevation and decreased pressure, the air cools slowly (adiabatically) by 1°C every 100 metres (328ft), known as the 'dry adiabatic lapse rate'. With further elevation, part of the water vapour contained therein condenses to form clouds and rain or snow. In doing so, it releases heat which, in turn, reduces the rate of cooling to 0·5°C per 100 metres, known as the 'wet adiabatic lapse rate'. After the air current has crossed the peaks and begins to descend, the clouds disappear, the air warms with compression and takes up moisture. The rate of warming is dry adiabatic from the outset, that is 1°C per 100 metres. With mountains reaching 3000 metres (9840ft), the following could apply:

adiabatic cooling 1°/100m (328ft) to 1500m (4920ft) = 15°C
from 1500m (4920ft), cloud formation and cooling 0·5°/100m (328ft) = 7·5°C
adiabatic warming valleywards = 30°C
temperature gain = 7·5°C

The origin of the air is not important, for the Föhn is not a warm southerly wind, rather gaining its warmth from the process explained.

Effects on weather. Observations on the Alpine summits show the Föhn to be turbulent and wet. On the southern side of the Alps, the weather is also cool and wet. From the Alpine Foreland a wall of cloud can be seen over the Alpine summits, the top of the cloud lying on the south side. The Föhn can blow for several days. In north of the Alps, its break-up is usually followed by an arrival of cold air, this air becoming dammed by the mountains. Elongated lens-like clouds, parallel to the mountains are typical of Föhn conditions. Such clouds are most striking in the deep blue skies.

Forecast. Föhn winds are not associated with a persistent weather system. In the westerly conditions more common in the Alps, the Föhn mechanism is lost, low pressure surviving in southern Germany. Generally, the cold air of the depression passes through the Foreland from the west, destroying any Föhn effect. A warm moist southerly air flow is most favourable to the development of the Föhn. With a well-marked southerly system, in conjunction with a stationary high over eastern Europe and a depression over the Atlantic, the weather is good and further improved by the Föhn mechanism.

Föhn effects have here led to the formation of wave-form altocumuli

Fronts

Recognition. The first signs of a cold front are soft cirrus clouds, followed from the west and northwest by banks of thick cloud. The wind freshens, and blows hard. Temperature falls and pressure rises. Rainstorms or snow showers come next and a rapid increase in temperature may well occur. The weather remains changeable. What we have then observed is the passage of a cold front.

The passage of a warm front is much more calm. Once again it is cirrus clouds that herald the arrival, followed by stratus clouds. Without much wind movement, gentle rain begins which then increases to a steady and fairly heavy rain belt. The barometer shows a fall in pressure and, after the front has passed, there is a sharp rise in temperature with a clearance of cloud cover.

Causes. Since cold air is heavier than warm air, it spreads at ground level while the warm air occupies higher levels. Meteorologists compare the distribution of warm and cold air masses at selected heights. The warm air originates in the south, the cold air comes from the north. In European latitudes, these two air masses are always opposed and moving, without stabilizing in relation to each other. Indeed, their frequent interchange usually results in changes in the weather. The contrast in temperature brings on turbulence, and stability can only be achieved when the two air masses are thoroughly mixed. It is at the frontal zone, where the two air masses are in contact, that the greatest changes in the weather occur. Cold air moving across warm stationary air is described as a 'cold' front, whereas, when the warm air pushes against stationary cold air, the term used is a 'warm' front. The moving cold air lifts the warm air, the moving warm air glides over the cold air.

Effects on weather. Cold and warm fronts are normally in close association with a depression (see p. 42). Since the horizontal and vertical extent of cold and warm air masses is extremely variable, the weather sequence is also variable. Again, the duration of cold air inbreaks and the movement of warm air over cold is extremely variable. Westerly weather conditions, possible in all seasons all over Europe, are typical: from west of Iceland the north wind drives cold air forward, while from the west of a high over the Azores the south wind drives warm air forward. Where the fronts meet, a series of eastward-moving cyclones develop. The weather becomes calm when the warm air sector finally lifts clear from the ground, and two cold air masses, with different temperatures, meet at ground level. This is known as an 'occlusion'. When a depression approaches a mountain, the rate of development of the occlusion is accelerated. The weather sequence is characterized by cloud masses, rainfall, wind and temperature change.

Forecast. Fronts are closely connected with cyclonic activity. With the passage of a cold or warm front, changeable weather can be expected at the observation point for a few days. Where the front pushes against mountains, for example a weather system in western Norway and in the Alps, a longer period of rain occurs.

Approaching cold front over the sea with a massive belt of cumulus cloud (*above*)

Roll of cloud over the sea, forming an air mass boundary accompanied by a wind change (*below*)

Hail

Recognition. Hail consists of ball-, egg- or pear-shaped ice particles, usually falling in a storm shower. Their diameter varies from 5 to 50mm (0·2 to 2in), although in very exceptional circumstances hail stones have been recorded with a diameter of 6 to 7cm (2·4 to 2·8in) and a weight of 1kg (2·2lb). Sleet is rain and hail intermingled, the hail usually being of small diameter. Hail falls from towering cumulus cloud, which reach great heights, with a spreading top or anvil, usually broken at the edges in a fibrous veil, showing a strong uplift and subject to icing. The cloud colour is a sulphurous yellow to grey-black. An actual hail storm is usually very sharply defined, and the fall rarely lasts more than a quarter of an hour.

Causes. A fluctuation of up and down draughts is present in the cumulus cloud that keeps the precipitation aloft. Because of the heights involved, the clouds contain ice particles and supercooled water droplets (see diagram p. 109). The hail gains weight as the constituent particles link up with each other. Falling ice particles strike the supercooled water droplets, which freeze around them either as pure ice or as a milky layer containing air. When the updraught is 30 metres (98ft) per second, the fall of these hail particles is stopped and they are kept aloft or pushed back to a higher level. More water droplets are impacted and the hailstones grow in size. This process continues for as much as an hour before the mass of the hail stones becomes too great for the updraughts to support. The way in which the hail stones are formed can be seen in their structure (concentric shells).

Effects on weather. Hot or frontal weather is generally necessary in bringing on hail and sleet showers, but ice particles can occur as precipitation when a belt of polar air reaches Europe in association with western and northern weather systems. In the latter case, however, the precipitation is usually in the form of sleet, as in March, when showers consist of sleet, hail and rain all mixed together. Not all sleet and hail reaches the surface in solid form. Melting can take place while the particles descend through the air, resulting in large to very large raindrops.

Because of the damage to crops, particularly to fruit and on occasion to buildings and vehicles, attempts have been made in areas most often experiencing hail to provide some defence by preventing hail stones from forming. Some of the methods used, such as cannon fire, have understandably proved useless, but a measure of success has been achieved by seeding clouds with silver iodide and other silver salts in the hope of preventing the continued growth of hail stones.

Forecast. In addition to the indicators described under storms (see p. 84), hail is often evidence of an imminent cold front. With the passing of the front, the wind veers, temperature falls, pressure rises and precipitation comes – not infrequently as sleet together with heavy rain. The following days are usually unsettled with a tendency to rain (or snow in cold seasons).

Rapidly-ascending cloud with hail showers

Halo phenomena

Recognition. Streaks and arches of colour sometimes form a ring around the sun and moon. Such rings with the sun or moon as central point generally have a dull red colour on the inner side, and often a violet rim on the outer side. Vertical columns of white light above or below the sun or moon, with flecks of light to the left or right, can also be observed – the so-called counter-sun and counter-moon. All these phenomena occur only in conjunction with high thin cloud, sometimes so thin that it appears non-existent.

Causes. 'Halo' has a Greek root, meaning circle or ring. These atmospheric halos are optical phenomena, they vary in appearance and are the result of refraction and reflection caused by ice crystals in the atmosphere. These ice crystals belong to cirrostratus cloud (see p. 86) which is usually so thin that the heavens are only lightly veiled. The ice is present as hexagonal crystals, thus forming a six-sided prism with a six-angled cross section and a smooth right-angled base. The path which sunlight follows through the ice crystals produces refraction angles of 22° and 46°. The rings around the sun and moon have therefore a radius of 22° and 46° (the smaller and the larger halo). The configuration depends upon the angle of the ice crystals in relation to the sun. The most important halos are:

type	dimension	frequency
small ring	22° radius	frequent
counter sun	22° to left and right of sun or moon	frequent, often with small ring
upper and lower tangent arcs to small ring arcs	exactly above or below sun or moon, 22° away	frequent, particularly the upper tangent, often very colourful
counter-sun circle (horizontal circle)	includes a white colourless streak, horizontal and same height as sun or moon	rare, usually in conjunction with 2 counter-suns 120° from sun or moon
light column	vertical strip of light above or below sun or moon	frequent with low or sinking suns, or shortly before dawn
large ring	46° radius	not so frequent as small ring, particularly colourful in the upper tangent arc
circumzenith circle	over sun or moon and 46° away	corresponds to upper tangent arc of large ring, very rare

Usually several of these features are observed together as one complex phenomenon.

Part of a halo

Effects on weather. Halo phenomena are linked with a thin high cirrus cloud cover (cirrostratus), pure ice clouds high in the troposphere. The extent of the cloud cover can be variable. Halos appear in association with thin wisps of cloud formed in high pressure conditions, and also in a phase after good weather, when cirrostratus forms a continuous cloud cover. They can appear at any time of year.

Forecast. Halos do not indicate any definite deterioration in the weather. Cirrus cloud is in no sense a rain cloud, and cirrus cloud cover is possible in good weather. Nevertheless, the cirrus clouds do indicate air movement high in the troposphere, particularly when cirrostratus forms rapidly. This is indicative of a change to alto-stratus (see p. 86) and the approach of a rain belt.

Haze

Recognition. In good weather, particularly when that good weather has lasted for a long time, a dirty yellow or red layer can be observed towards the horizon which, against a dark background, takes on a bluish colour. Even directly above, the blue skies become covered with a grey white veil. Both the distant haze and the thin veil herald dull weather, and this dullness is most marked in association with high pressure and an inversion. Visibility is poor, and in urban and industrial areas, even in cloudless weather, the passage of sunlight is hampered. As the haze thickens, visibility decreases and mist develops with grey tones predominating. In fact, haze, when following rain, is often the precursor of fog.

Causes. Haze may be either dry or wet. Dry haze is a mixture of dust, smoke, sand and dirt particles whisked from the earth's surface into the atmosphere. The dullness associated with dry haze can be due to some local occurrence, such as a sandstorm, or the result of industrial pollution or of a forest fire. Generally, the haze is caused by particles from a variety of sources. What is an important fact is that these heavy particles in the air act as condensation nuclei for precipitation. If various preconditions are met, the water vapour in the atmosphere condenses on the heavy particles and precipitation follows. Lack of visibility may also be the result of a great density of minute water droplets in the atmosphere. A moist haze can quickly thicken into a mist. A small drop in temperature is usually sufficient to make the water droplets grow. When visibility remains over 1000 metres (3280ft), the term haze is still used, but, if less, it is known as mist.

Effects on weather. A dry haze, usually rich in water vapour, generally develops during good weather conditions. A relatively light wind is very often sufficient to spread dust particles through the lower atmosphere, helping to bring on what is normally described as heat haze. The situation is, however, quite different when, in conjunction with a depression, moist air cools and begins to form a veil of haze. This is the beginning of mist or cloud development.

Forecast. The nature of the haze and its density are important to prognosis. If the air is relatively dry, and the haze itself therefore dry, a further loss of visibility in no way indicates the break-up of high pressure conditions. Particular attention must be paid to a loss of visibility following an influx of warm moist oceanic air. The increased haziness and development of overcast conditions, often accompanied by the formation of stratus clouds (see p. 86), is evidence of a weather change, generally heralding rain. In distinguishing the nature of the haze, twilight colours are significant (see p. 40).

Typical haze in mountains, with limited visibility (*above*)

Haze over the sea. Stratocumulus clouds further above (*below*)

High pressure areas

Recognition. Fine weather with blue skies and isolated short-lived cumulus clouds with little or no wind are associated with establishing high pressure in summer. If the weather persists, with greater temperatures, a heat wave results. In winter, however, high pressures can lead to severe frost and, instead of the blue skies, high altitude mist or stratus cloud and inversion conditions (see p. 62). Short periods of clearer weather, after a depression has passed, give good visibility and isolated showers, but such clear periods last only one or, at the most, two days.

Causes. The disappearance of cloud is caused by the sinking movement of air. Because of this lack of cloud, the spiral movement of winds in high pressure areas is not revealed in satellite photographs. In the northern hemisphere, this wind movement is clockwise (anticyclonal). Usually the air mass of a high pressure area is uniform. The wind flows from the high pressure towards low pressure areas, although not directly because of friction with the earth's surface. With this outward movement, air from above sinks and, in sinking, warms – with the consequent absorption of the clouds. When the air in the high pressure area is warm, the temperature drop at higher altitude is less than in a depression; nevertheless, in summer the air near the ground is warm, in winter it is cold. The build-up of high pressure is usually slow. In the subtropics, there are permanent high pressure areas which, because of the air circulation in these latitudes, are zones of sinking air (see air pressure maps, p. 24). The northern subtropical high pressure zone is of great importance to the weather of Europe, influenced by warm air flows from the region centring on the Azores. Extending from this area are wedges of high pressure between depressions. The trigger for this extension of high pressure is the cold air to the rear of the eastward-moving depressions. The movement of the cold air towards the high pressure area is slight, and so is its influence on the weather sequence.

Effects on weather. Easterly and southerly weather conditions are characteristic in the establishment of high pressure areas over large parts of Europe, together with a stationary high over Scandinavia. Since the sinking air facilitates cloudless conditions, good weather can be anticipated – but there are disadvantages. The intense heating of the ground in summer should bring on cumulus cloud and summer thunderstorms (see p. 84). In winter there is a radiation deficiency because of the long nights, a context which promotes fog. The Azores high, as mentioned, is of particular importance to European weather, both for producing good weather and constantly-recurring westerly conditions. In contrast to the stationary high areas with long periods of good weather, the high pressure wedges from the Azores high are very mobile and have only a short-term influence on the weather. The wedges bring one or two days of calmer, clearer weather between depressions, with little wind and only a few showers.

Forecast. So long as the cumulus cloud lasts no longer than a few hours (see p. 38), no deterioration should be expected. Again, increasing mist and fog in winter or autumn are not indicative of change. On the other hand, the appearance of cirrus clouds can be significant.

The appearance of a few cumulus clouds in the morning and about midday – not unusual in a high pressure area

Humidity

Recognition. Rain, hail and snow are not the only forms of precipitation. Moisture can be present in many forms, for example, on a summer evening when the grass is suddenly wet with dew, or near the coast as moist sea air drifts onshore. Inland, as well, there are variations in humidity – the air may well feel either dry or moist. A noticeable instance of precipitation is the condensation on windows. When warm air meets the cold glass, the moisture collects in droplets and runs down the panes.

Causes. Air without exception contains moisture as water vapour – sometimes more, sometimes less. Water vapour cannot be seen, smelt or tasted. For a given temperature, the air holds only a fixed amount of water vapour. The warmer the air, the more moisture it can hold; the cooler the air, the less. For each cubic metre, the levels are:

water vapour (grams/oz)	temperature (0°C)	water vapour (grams/oz)	temperature (0°C)
30·3/1·069	+30	4·8/0·169	0
17.3/0·610	+20	2·4/0·085	−10
9·4/0·332	+10	1·1/0·039	−20

Although temperature is thus important, it does not indicate directly the amount of moisture present. It is also necessary to know the relation of absolute humidity to saturation point at that temperature. The result is relative humidity. For example, in air at 20°C with 10 grams water vapour per cubic metre, the relative humidity is:

$$\frac{10 \times 100}{17 \cdot 3} \qquad \frac{\text{absolute humidity} \times 100}{\text{saturation humidity}}$$

The result is approximately 60 per cent relative humidity. If this cubic metre of air is cooled to 10°C, the saturation humidity falls to 9·4 grams, and 0·6 grams is released. as dew in warm weather and as hoarfrost in cold weather. There is always a point at which an air mass becomes saturated – the dew point.

Effects on weather. A knowledge of air moisture and its response to warming or cooling is extremely important. Warmed air retains its water vapour but the relative humidity falls, the air becomes relatively drier. Cooling of air leads to an increase in relative humidity up to the saturation point. Beyond that point, condensation or sublimation (the formation of ice crystals) takes place, and either dew or hoar frost forms on the ground.

Forecast. Warm air with a relatively high moisture content enhances the development of clouds and rain. The colder the air, the lower its relative humidity, the less the likelihood of clouds and rainfall.

Hoarfrost on trees and other plants are a result of a transition from water vapour directly to ice (sublimation).

Ice

Recognition. Ice can be seen almost everywhere in western Europe in several forms. From the atmosphere come such crystalline forms as sleet and hail, while frozen soil moisture occurs as black ice and ground frost; and then there is ice formed on open water. Icing of the soil can often be seen after a snowfall (see p. 80) and hail. Ground in the shadow of forests or mountains is particularly vulnerable to icing, for example, on forest roads. Cold winds can also lead to icing, and, of course, season and latitude are significant. Sea ice forms closed ice sheets as the temperature falls, broken and compressed by wind and sea into pack ice. Drift ice is formed by the melting and break up of the ice sheet.

Causes. Ice is water in a solid state. Freezing takes place at 0°C at a pressure of 1 atmosphere. With falling temperatures, the density of water does not continuously increase; indeed it decreases below 4°C, the temperature at which maximum density is reached. At freezing point, the density has diminished by 10 per cent, but ice-like structures are formed before freezing. The reduction in density explains why ice floats ('the water anomaly'), and why water freezes from above. Air reaches the freezing point of water by vertical movement and by cooling. The contained moisture is then frozen out. The level at which freezing occurs is not constant because of the horizontal movement of warm air. On average, temperature falls by 0·5°C per 100 metres (328ft).

Effects on weather. When the temperature drops below 0°C, the weather is described as being frosty. Such weather is not restricted to winter. Reference is made to a frost day if the lowest temperature is below 0°C, and an ice day if the temperature does not rise above 0°C throughout the day.

The following should be distinguished: *radiation frost* follows the loss of heat by radiation on clear still nights, for example night frost in spring and autumn; *advection frost* is the result of an influx of cold air, for example polar air with northern weather conditions early in the year; *ground frost* is frost which forms on the ground and up to 2 metres (6·5ft), mostly as the result of radiation.

In much of west and central Europe, soil water freezes down to a maximum of about one metre (3·28ft) while in Siberia and Alaska some of the ground never thaws (this phenomenon is known as permafrost). Volume changes as a result of freezing cause frost damage and frost wedges in the soil. Radiation frost occurs mainly in high pressure conditions, but advection frost is associated with an inbreak of cold arctic air and with showery weather, most often in March. It also occurs in good weather, for example, in easterly conditions with a stationary high over Scandinavia and a depression over the Mediterranean.

Forecast. Clear still weather with radiation frost is indicative of stability, particularly in autumn and winter. Inbreaks of cold air are less persistent, but in the mountains they bring ice and snow, especially in winter. In central Europe, black ice shows the transition from frosty weather to thaw as the warmth increases (as in a warm front combining with a westerly airflow).

Ice on a stream, with snow sheets, and with rime on the trees and bushes

Inversion

Recognition. In clear weather, particularly in autumn and in winter, a dirty haze can often be seen that visibly flattens off at its upper limits. These upper limits are highlighted by the presence of rising smoke that here flattens and spreads laterally, giving the impression of no further rise. Visibility is very limited. In mountain districts, a similar high-lying sheet of haze or mist can be observed, also clearly delineated at its upper limits. All vertical air movement appears to be forestalled.

Causes. Clear weather in autumn and winter promotes radiation loss at night. The layers of air near the ground cool quickly so that cold air lies close to the ground with warm air above. Mist forms in the cold air. Temperature measurements show a departure from the normal fall of temperature with height, with increasing temperature to the top of the cold air. The normal fall in temperature is resumed from the cold air/warm air boundary – the inversion level. This stratification is astonishingly stable. The inversion level acts as a barrier to the rise of warm air and to the rise of smoke and exhaust gases. Vertical heat exchange becomes exceedingly difficult. In industrial and residential areas, smoke and fumes accumulate beneath an inversion and lead to marked air pollution that eventually can be very dangerous. In autumn and winter, the daily warming is not sufficient to produce upward movement through the inversion layers.

Inversions can also develop at high levels as a result of the rise in pressure after a depression and a groundwards-directed air stream. A 'permanent' inversion is present at the tropopause – the upper boundary of the troposphere.

Effects on weather. Inversions are linked in particular with autumn and winter high pressure systems, for example, southerly weather conditions (highs over Scandinavia, depressions over the Mediterranean) and easterly weather conditions (highs over eastern Europe, lows over western Europe). The low elevation of the sun makes it difficult to break up the inversion. Weather in October and November frequently features this condition. The inhibited vertical movement of air is resumed if the high pressure weakens. This is usually the case if strong winds blow up with the establishment of westerly conditions characterized by changeable weather. The air layers up to the inversion layer are then replaced, and the pollutants above towns and industrial areas blown away.

Forecast. The vertical rise of smoke marks a normal temperature lapse rate in the atmosphere and is characteristic of high pressure. Nevertheless, deflection at the boundary of an inversion layer is also indicative of high pressure, an inverted lapse rate. Such conditions can be very persistent. At higher levels, above the inversion, it can be sunny and relatively warm, while below in the valleys it is very misty. No sudden change in weather can be expected.

Inversion with haze beneath (*above*)

Smog with high chimneys emitting above the fog layer (*below*)

Lightning

Recognition. Lightning is an electrical spark discharge usually occurring in a storm (see p. 84). The track made by the lightning can vary in length and complexity: it can be linear, or branch out extensively from above or below (forked), an illumination of the clouds (sheet) or a fire ball.

Causes. Every lightning flash is an electrical discharge, made necessary by a strong electrical field formed in storm clouds, between positively and negatively charged parts of the cloud. The potential can be built up in the clouds, between clouds, between clouds and the earth's surface, or between clouds and the upper levels of the atmosphere. Usually there is an initial weak discharge, followed by the main discharge, which can be multiple. The potential can reach 100 million volts. The energy involved in each lightning flash is estimated at on average 100 kilowatt hours. The loss of potential occurs in a very short time (on average 50 seconds), so that the electrical charge is up to 100,000 amperes. All lightning follows a track: the most common form is forked; sheet is not directly visible, its appearance as an illumination of the clouds probably being caused by its occurring beyond the horizon; ball is still not really understood but may be an electrical dust cloud, a long-burning mixture of air and carbon, or an ionized gas mass formed in a lightning track.

When lightning strikes the earth a downward discharge develops millions of volts in about one-millionth of a second, cutting a path through the air. Then a return charge flashes back up from the ground, lasting from 5- to 200-millionths of a second. Finally, a burning, melting charge strikes back to the ground, lasting up to one-tenth of a second. The number of lightning strikes is very small in proportion to the total number of lightning flashes. They occur as discharges from cloud to earth, and vice versa. If any conflagration takes place, it is the result of lightning with a low charge, but a long discharge. Very brief flashes are cold strikes since, although temperatures of $10,000°C$ are reached, these are too short-lived to fire the material involved.

Thunder is probably caused by a compression wave set up by the expansion of air, through which the lightning has passed, and by the collision of the air particles returning to fill the vacuum which has as a result been created. Since sound travels more slowly than light, the interval between seeing the flash and hearing the thunder is an indication of the distance of the storm's occurrence: a 10-second gap would indicate about 3km (1·85mi).

Effects on weather. Lightning accompanies both summer and frontal storms. With very distant storms, thunder is heard but lightning observed only as illumination on the horizon. The heavy rain or hail normally associated with lightning is caused when moisture is forced upwards rapidly in a thunderstorm, with rapid freezing and precipitation ensuing (see p. 84). Sometimes, so much heat is produced that a dry storm occurs, the rain being re-absorbed in its passage through the warm air masses.

Where lightning will strike is in part decided by the conductivity of the earth's surface: moist ground is more dangerous than dry ground; trees and exposed buildings are vulnerable.

Forecast. Strong lightning is usually accompanied by heavy rain. Lightning seen as an illumination of clouds in the west or southwest indicates an approaching front. Similar illumination to the east or, in most cases, northeast means that no storm should be expected. Distant lightning during a high pressure situation may end with a local storm, but does not herald a complete weather change. Indeed, the release of energy in a thunderstorm tends to stabilize a weather system.

Forked lightning

Mist and fog

Recognition. Mist and fog form white cloud-like swathes and veils hanging thickly over the ground or at low elevations, particularly in the evening, at night and in the morning. Haze is often a precursor of mist, and frost is often associated with fog as hoarfrost or ice. With a thick fog, visibility can be reduced to 5 to 10 metres (16 to 32ft).

Causes. Fog and mist are cloud forms developed at the earth's surface. They originate out of warm moist air mixing with cold air (mixing fog) or because warm moist air comes into contact with the cooled earth's surface (radiation fog). In either case, the cooling process leads to condensation, but is not so strong as to lead to rainfall. Occasionally however, a fog can be accompanied by a very fine rainfall, the so-called 'foggy rain'.

Effects on weather. Fogs are associated with particular daily and seasonal weather conditions, see chart on p. 68.

Radiation fog (typically, ground fog) is associated with calm and clear skies (high pressure system). The mixing fogs (typically, coastal fog) occur in overcast conditions and are blown by the wind, particularly westerly winds.

In addition to rain and drizzle, wet fog is one of the prerequisites for the development of 'black' ice. The fog particles freeze on contact with the ground. If the temperature of the droplets is more than $0°C$, the ground's temperature must be even lower. If the droplets are supercooled, 'black' ice can form even though the ground is at first a little above freezing point. These conditions are particularly dangerous for road traffic, usually occurring at the change of the seasons between autumn and winter and between winter and spring.

Forecast. Fog can occur in conjunction with various weather conditions, the view that lifting fog is accompanied by bad weather being somewhat too simplistic. The disappearance of fog is closely linked with solar radiation, required for the warming. With radiation fog, incoming radiation is unhindered and the fog usually lifts in the course of the morning. If the fog is very thick, say 200–300m (650–980ft), the warming by the sun may prove inadequate. The build-up of stratus clouds can block the sunshine and sustain the fog. The appearance of the stratus cloud may, however, herald a deterioration in the weather. In every case, the fog, whether or not it lifts or falls, will ultimately be dissolved by increasing warmth from the ground.

Surface mist in a mountain foreland (*above*)

Fog at sea (*below*)

Mist and fog

fog type	development location	development time	visibility	associated conditions
lake fog	water surface	evening, night and early morning in spring, winter and autumn	10–20m (33–65ft)	at freezing point, glazed frost and rime
ground fog	flat land and mountain valleys	evening, night and early morning in autumn, winter and spring	5–10m (16–33ft)	develops suddenly and in winter may be persistent
wet fog	all types of terrain, especially roads	after rain, sun heats the ground	15–20m (49–65ft)	usually only a couple of metres thick
river fog	water courses	late summer to mid-winter in evenings, nights and early morning	10–20m (33–65ft)	particularly intense in November, December, January
frost fog	meadows and arable land	in winter with deep frost	10–20m (33–65ft)	'dry' fog, not causing icing
storm fog	woods	in summer after heavy rain	15–20m (49–65ft)	occurs suddenly locally limited
coastal fog	sea and coastal area	all year round	5–10m (16–33ft)	mixing fog, carried inland by onshore winds
marsh fog	moist ground marsh	all year round; little relation to time of day or weather conditions	5–10m (16–33ft)	often only a couple of metres high
smog	urban areas	particularly in autumn and winter, may be persistent	1–5m (3–16ft)	fog polluted with smoke and dust particles
valley fog	basins, hills and troughs in mountains	evening, night and morning; all seasons, moist still conditions	10–20m (33–65ft)	rising shreds of mist are indicative of development
mountain mist	mountains and plateaus	all seasons cool, windy weather	5–10m (16–33ft)	mixing fog usually wet and in winter associated with 'black' ice

Thick ground fog in the Alps at 954 metres (2800ft) above sea level

Moonlight

Recognition. In clear weather it is possible to see the four phases of the moon. The complete cycle takes four weeks, beginning with a thin sickle in the western sky. The size of the moon increases (waxes) to the first quarter (half moon) and finally to the full moon. At this stage, it is visible throughout the night. As it decreases (wanes), it shines for shorter and shorter periods. Because of the purity of its light, both the waxing and waning phases can often be seen during the day. At the time of the new moon it is also possible to see 'the old moon in the new moon's arms', the dark part of the moon being lit palely by reflected light from the earth. In thin cloud cover the moon can be seen with haloes and coronae. If the full moon passes into the shadow of the earth, then it appears in eclipse and takes on a characteristic copper-red colour.

Causes. The moon is the earth's nearest neighbour. It takes four weeks to circle this planet and its phases depend upon the relative position of earth, moon and sun. A new moon occurs when all three are in a straight line with the moon between the earth and the sun. The first and last quarters occur when the moon and the earth are positioned at right angles to the sun-earth alignment. The full moon takes place when the three bodies are again in a straight line, the earth being between sun and moon. The phases show that the light from the moon is a reflection of radiant energy from the sun. Since the rotation period of the moon is the same as the time it takes to circle the earth, the same side of the moon is always that which is visible to any one part of the earth. The moon's most important effect on the earth is its influence on the tides, the rhythmic oscillation of the ocean surface, and also of the surface of the earth and air pressure. Tides are caused by the gravitational pull of the heavenly bodies, in particular, of the sun and the moon, which periodically disturb the gravitational field of the earth.

The moon has a diameter of 3470km (2155mi), and is on average 384,400km (238,700mi) from the earth, the distance varying from 363,300km (225,600mi) to 405,500km (251,800mi).

Effects on weather. The direct influence of the moon on weather conditions has yet to be satisfactorily explained. The often-quoted connection between the waxing or full moon and clear skies is based on a false premise. Clouds do not clear because the moon shines, rather, the moon is visible because the sky is clear. A waxing moon is more likely to be seen because it shines in the early part of the night. The same is true of the relationship between the moon and frost. Clear weather in winter is one of the causes of night frost and, of course, with no cloud the moon is visible. Furthermore, the moon's tidal attraction is too small to produce significantly strong changes in air pressure to influence weather conditions.

Forecast. The popular theories that a waxing moon means good weather, a full moon heralds a change in weather and a waning moon means bad weather are without foundation. There is no direct relationship between weather conditions and the moon alone. However, optical phenomena associated with the moon, such as haloes and coronae, give significant information about conditions (see pages 52 and 36). The colour of moonlight depends on atmospheric conditions (see p. 40) and, from this, weather changes can also be forecast.

Moon with Föhn cloud

Noctilucent cloud

Recognition. These are relatively rare phenomena, observed in the northern heavens in summer months for one to two hours after sunset. Noctilucent clouds appear similar to cirrus clouds, fibrous, long drawn-out and silver-white in colour, often with soft blue tones. They frequently have a wave-like band structure and are so thin that the stars shine through. When they do occur, they may last for hours. Noctilucent clouds are not to be confused with the mother-of-pearl clouds, also observed after sunset but at lesser heights of about 30km (19mi). Observation of the noctilucent clouds is limited to latitudes 45°–70°N and S. The best period for observation is at the end of June and beginning of July.

Still higher than the notilucent clouds are the so-called 'light streaks' with an average altitude of 125km (78mi). They are less distinct than noctilucent clouds and are most often observed from October to February, rarely seen in the months March to September. Seen from southern England, they appear as features of the northern heavens. Their origins are now thought to be cosmic dust which has entered the earth's atmosphere.

Causes. Noctilucent clouds are distinguished from all other clouds by their great height, which ranges about 70 to 90km (43 to 56mi). Their first recorded observation was in 1885, two years after the Krakatoa volcanic explosion, and they were singled out for their extraordinary height. Their light is, in fact, scattered sunlight. The sun must be at least 6° below the horizon and the clouds are then observed against a light sky. Although they last for hours, individual noctilucent clouds show rapid structural alterations. Their frequency of occurrence is extremely variable. In order to appreciate their structure, it must be remembered that the density and pressure of the atmosphere at these heights is 1/100,000th of that at ground level. It is indeed remarkable that there is still enough water vapour for condensation to take place. As with clouds at lower levels, whether of ice or water droplets, condensation nucleii must be present (see p. 54). It has been speculated that, in the case of noctilucent clouds, the nucleii are provided by cosmic dust. In order to maintain the supply of water vapour, air must be carried up from the lower levels of the atmosphere.

This carrying of water vapour as the result of volcanic eruptions or nuclear weapon testing in the stratosphere has not been proved. If volcanic eruptions do project water vapour above the stratosphere, there should be a correspondence year by year between the frequency of volcanic outbreaks and noctilucent clouds. This is not the case. The American specialist, Banhard Haurwitz, who has studied noctilucent clouds for years, is of the opinion that considerable atmospheric movement is required for the transport of the water vapour, but that direct observations by rocket probes would be needed to resolve the problem.

Effects on weather. Although it is now known that noctilucent clouds are not dust but ice with condensation nucleii and are thus related to cirrus clouds, they have little significance for the weather. The same also applies to 'mother-of-pearl' clouds.

Forecast. Noctilucent clouds occur with abnormal twilight conditions and were once thought to indicate the presence of dust in the higher atmosphere. Recent investigations of the volcanic theory, however, have proved that theory to be untenable.

Noctilucent cloud 7–8 June, 1976 (photograph taken at 01.00 Greenwich Mean Time)

Northern lights (Aurora borealis)

Recognition. Without question, this is the most beautiful heavenly phenomenon but, unfortunately, it rarely occurs. It can take the form of waves of light rolling from the horizon to the zenith, or can appear as arches of colour over the horizon with many individual structures, such as rays, penetrating in all directions – the streamers or Merry Dancers. Red colours are present throughout but, in higher latitudes, greens and yellows are dominant, with flickering tints of blue and silver. The region in which they may be seen most often is in Europe near the Arctic Circle (Iceland, northern Norway, Lapland). In these far northern latitudes, the Aurora may be seen on an average of 100 nights a year, whereas in northern Scotland the frequency drops to thirty, southern England five, southern Germany one, and in southern Italy the northern lights are seen once in ten years. The intensity is extremely variable. It can be strong enough to light the heavens as brightly as a full moon, and to make reading a newspaper possible.

Causes. The Northern lights are optical phenomena of the upper atmosphere, of great variability in colour, brightness and form. They occur generally in the ionosphere at heights of 100 to 400km (60 to 250mi). They have been recorded at 70km (43mi) and also at as much as 1000km (620mi), but peak occurrence is in the 100–150km (60–90mi) range. Scientists have established that the occurrences follow an eleven-year cycle, parallel to the development of sun spots. It is now thought that the Northern lights are due to the entry into the atmosphere of electrons and protons from the sun. On penetrating the atmosphere, these so disturb the atmospheric oxygen and nitrogen that light is emitted. The solar particles, which are already up to speeds of 1000–2000km (621–1242mi) a second (corresponding to a journey time from the sun of twenty to forty days), accelerate on entering the atmosphere. The magnetic field of the earth causes these particles to penetrate furthest at or near the magnetic poles. It is then that the ionization processes and electrical flows take place, with consequent disturbance to short-wave radio transmissions.

Effects on weather. The level of the atmosphere at which the Northern lights occur, is the ionosphere. Its base is about 60km (37mi) from the surface of the earth, its upper limits 2000km (1242mi) and it reaches the outer layers of the atmosphere. At 50km (31mi) altitude, air pressure is a thousandth of its average at ground level, about one millibar (mb). At this point, it should be noted that there is no form of weather in this very thin atmosphere. Weather events of daily significance are confined to the lower atmosphere at a maximum height of 15km (9.3mi). Despite this, scientists argue that this stream of solar particles cannot be ignored in terms of possible effects on weather. Relationships have been suggested between the stream of particles the existence of which is made known by the Northern lights, and the development of extensive depressions over the northern Pacific Ocean.

The Northern lights produce a brilliant display of colours

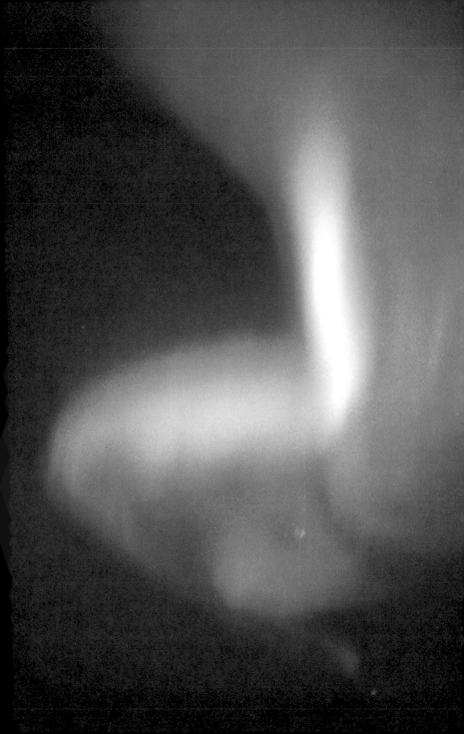

Rain

Recognition. Precipitation in the form of water droplets occurs at various degrees of heaviness. The finest rain, drizzle, often accompanies mist. Light rain, usually in the form of showers, consists of small droplets with a corresponding lack of speed in falling. Heavy rain is often frontal, and the heaviest, usually described as 'cloud burst' is normally associated with a storm. The rain drops are very large and fall fast, bringing astonishingly large volumes of water in a very short space of time. In colder periods of the year, particularly those marking the beginnings of autumn and of spring, rain is sometimes observed that freezes on impact, producing icy conditions.

Causes. Air contains water vapour, and the cooling of air leads to the condensation of that vapour. This is the result of the fact that air at a given temperature contains only a definite quantity of water vapour. At 20°C a cubic metre of air contains 17.3 grams (0.61oz) of water vapour, at 0°C only 4.8grams (0.17oz). The condensed water vapour falls from the clouds to the ground as rain, unless it is again evaporated en route. It is important for the formation of raindrops that the water droplets or ice crystals grow until they are heavy enough to fall. The joining together of several droplets is influenced by the movements in the air and by the electrical conditions of the clouds. A normal rain drop of a few millimetres diameter is composed of a million of the smallest water droplets.

rain	diameter of droplet	fall speed
drizzle	0.006–0.06mm (0.0002–0.002in)	0.10–20cm (0.039–7.87in/sec)
light showers	0.06–0.6mm (0.002–0.02in)	20–100cm 7.87–39.37in/sec)
heavy rain	1–3mm (0.04–0.12in)	150–400cm (59.06–157.48in/sec
cloud burst	4–6mm (0.16–0.24in)	500–800cm (196.85–314.96in/sec)

Depth of rain is the quantity falling in a given time (per day, month, year). A day having at least 0.1mm (0.004in) is called a rain day.

Effects on weather. Mobile warm air rises over static cold air and forms stratus clouds which bring rain. This happens, for example, when warm moist sea air moves inland (westerly weather conditions). Conversely, the inbreak of cold air into the static warm air brings about cumulus clouds and rain showers, the cold air coming from either west or north (northerly weather conditions). With a cooler system, the precipitation changes to snow. Warm summer storms bring heavy rain. In many areas, mountains and high ground cause rain that can persist for days, the moving air being dammed back and lifted. If the air movements are predominantly in one direction, then the windward side of the mountains will have heavy rain, and the lee side lie in a rain shadow. This is true of Norway, where the west coast is very wet, while the southeast is relatively dry.

Forecast. The heavier the rain, the larger the drops are and normally the shorter the duration. The approach of low-hanging clouds after a pause in the rain does not herald a coming change in the weather. The following air is cold and moist, and will not lead to a break in the cloud cover.

Rain air during the passage of cold front (*above*)

Nimbostratus clouds, with rain falling (*below*)

Rainbow

Recognition. This arch, lit with all the colours of the spectrum, the radius at an angle of 42° to the eye, is formed when the sun is behind the observer shining through a rain cloud in front of the observer. Violet is the innermost colour, red the outermost. Often two rainbows are seen, the secondary rainbow lying outside the primary, having a radius of 51° and being paler. The colours are reversed with violet outermost, red innermost. Occasionally, up to six further rainbows are observed in two series, one series beside the primary, touching the inner violet, the other series outside the secondary bow. From high mountains and from aircraft the bow can be seen as a full circle.

Rainbows are also seen in conjunction with moonlight, but are then white with a pale red rim.

Causes. The rainbow is an optical phenomenon of the atmosphere. It does not occur if the sky is either completely clear or completely overcast. A prerequisite is a local area of rain through which the sun shines, although the effect can be produced by the sun shining through a waterfall or fountain. The rainbow is formed as a result of refraction and reflection by the individual water droplets. The refraction takes place as the light enters and leaves the droplet, dividing the white light into the colours of the spectrum. These reach the observer by reflection. The individual droplets are not distinguished and the rainbow appears as a continuous band. The first reflection produces the primary bow, the second reflection the secondary bow. The intensity of the colour and the width of the rainbow depend upon the size of the drops. Drizzle (water droplets 0.05mm) produces only a very pale, almost colourless rainbow.

Effects on weather. The larger the water droplets, the more intensely coloured is the rainbow. The most beautiful rainbows occur in conjunction with heavy showers, as for example during a storm. The heavy rain produced by westerly weather conditions consists of drops sufficiently large to produce rainbows. The brief clear periods ensure that the sun's rays frequently fall on heavy rain. Often the rainbow is only partly formed because, as the clouds part, the rain is unevenly distributed. Snowfall does not produce a rainbow. The best conditions for rainbows to occur are during spring and autumn. The higher the sun is in the heavens, the more the occurrence of rainbows is limited to early morning and late afternoon.

An almost white rainbow develops in mist or together with fog. It is seen in the morning before the sun's radiant energy warms the ground and dissolves the mist.

Forecast. In the morning the rainbow is in the west, in evening it is in the east. Observed before midday, the rainbow shows oncoming showers from the west. No improvement in weather can be expected in the short term. Rainbows observed in the afternoon show rain clouds in the east, in many cases departing, for example, after a storm. Even in westerly weather conditions, a short-lived improvement is not ruled out as, for example, when associated with a wedge of high pressure.

A simple rainbow

Snow

Recognition. Snow occurs most frequently when the temperature is below 0°C. When the temperature is about freezing point, blizzards often take place with typical snowflakes but, at lower temperatures, the snow falls as individual crystals with a great variety of forms. At even lower temperatures, snowfalls are less frequent or absent. Snow also falls at temperatures above 0°C, mostly as showers, and even as high as 8°C, primarily in spring and autumn. In winter sports centres, a good deal of attention is paid to the quality of snow, distinguishing the fine-grain, dry 'dust' snow from the coarse-grained wet 'papp' snow, the latter occurring particularly at temperatures above 0°C. Fallen snow becomes consolidated (the formation of 'firn'). The wind moves the unconsolidated snow into drifts, some of considerable dimensions. In the mountains, wind-driven snow collects in hollows and builds the dangerous snow beds which form the sources of avalanches.

Causes. Snow forms in the air by the freezing of supercooled water droplets at temperatures of -12°C to -16°C. Crystals of various shapes such as needles, prisms, stars and plates are produced with a radius of from 0.005mm to a few millimetres, and a thickness of about one-thousandth of the diameter. The saturation vapour pressure (see Air Humidity, p. 106), as well as the temperature, influence both form and size. At about freezing point, the snow crystals flocculate (form into small wool-like masses, hence snowflakes). Since at very low temperatures the air has little moisture, the heaviest snowfalls occur with temperatures of about 0°C. For this reason, heavy snow is associated with middle latitudes and not with the arctic or antarctic where temperatures well below zero are experienced. The snowline of northern Europe, Asia and North America is the boundary between snow-covered areas and snow-free areas. The location of the snowline is dependent upon the altitude of an area and the conditions governing precipitation. Examples in Europe of permanent snow-covered areas (with the formation of glaciers) are:

area	altitude (lower limit)
Iceland	700m (2300ft)
Scandinavia	1300m (4250ft)
Spitzbergen	300m (1000ft)
Alps, north side	2500m (8200ft)
Central Alps	2900m (9500ft)
Alps, south side	2700m (8850ft)
Pyrenees	2600m (8550ft)

Effects on weather. If the temperature at the surface falls to 0°C, then snow falls instead of rain. The weather sequence is basically the same as for rain (see p. 76). It is notable, however, that the heaviest snowfalls occur during the passage of a warm front associated with rising air, called an 'anafront' (see p. 48). In the mountains this snow may lie, but in the lowlands the following warm air – warmer because it is nearer to ground level – usually brings a thaw. Northern weather conditions which bring showers to the lowlands in spring, bring snow in winter. Arctic air leads to cold air inbreaks, and frontal storms associated with cold air bring blizzards.

Forecast. With severe cold, about −5°C to −10°C, snowfalls are rare. Such precipitation as does fall, comes in the form of very fine snow crystals. In the lowlands, if the snowfall was the result of a well-marked westerly weather system, heavy snow is usually followed by a thaw. With extensive snowfalls, the air warms very slowly since evaporation of the snow requires a great deal of energy.

Starlight

Recognition. As darkness falls the heavenly bodies gradually become visible. Individual fixed stars, groups of stars in the Milky Way, the moon when visible (see p. 70), and some of the planets are the most important objects which can be seen without a telescope. The planets and the moon only reflect light from the sun, but stars all produce their own. The extent to which the heavens are visible depends upon the lack of sunlight (see twilight p. 40), and upon the transparency of the atmosphere. Water vapour, mist and air disturbances all reduce the visibility of stars, and an overcast sky will blot them out completely. The visibility of a bright star cluster provides a ready means of verifying the amount of moisture in the atmosphere and the twinkling of bright stars is indicative of atmospheric turbulence.

Causes. Local differences in the density and temperature of the atmosphere cause the twinkling of stars. If the atmosphere is particularly disturbed the twinkling can easily be observed with the naked eye and is particularly striking when seen through a telescope. The degree of disturbance of the atmosphere usually appears greater towards the horizon than at the zenith, but it also varies with time of day, being greatest at midday and least just before dawn and just after sunset. The light from the stars has to penetrate the atmosphere and is therefore subject to the turbulent movements within the atmosphere. These short-term changes in direction and brightness suffered by starlight, are responsible for the irregular-flickering – or twinkling. Starlight is also subject to absorption, so that not all the radiation from the star can penetrate equally the earth's atmosphere. In this case, the term extinction is used. Finally, starlight is also subject to refraction. Both extinction and refraction increase towards the horizon. Atmospheric moisture is particularly important in extinction, in particular over industrial or residential areas.

Effects on weather. Air movements are stronger in depressions than in association with high pressure systems but with greater cloudiness. It is often possible nevertheless to see the heavens during a weather depression since, in Europe, it may include short clear periods. The twinkling of stars during such clear periods is particularly marked, despite the tendency of the high moisture content to mitigate the disturbed condition of the air. This is true also of mist or fog, so long as the stars are still visible. Air is least disturbed during a high pressure situation, but even so very disturbed conditions can apply when the high is becoming established and as it breaks up.

Forecast. The twinkling of the stars betrays turbulent processes in the atmosphere. If the twinkling occurs after a few days of good weather, a change in air stratification is taking place, initially at great heights. If the twinkling is accompanied by poor transparency, caused by dust in the atmosphere, this can be taken as further evidence for a weather change. Often the turbulence in the atmosphere is associated with an increase in wind activity in the lower levels of the atmosphere. Unwavering starlight and good transparency show a stable weather situation, leading to frost in the cold season.

A small part of the Milky Way

Storms

Recognition. During the summer months, in particular, cumulus clouds (generally coming from the southwest) can suddenly expand upwards in towering masses. When these cloud towers spread out laterally and develop an anvil – a dark fibrous flat top – conditions are right for rain. Strong squally winds mark the beginning of the storm, together with heavy rain, or occasionally sleet, hail or snow. This precipitation is linked with the build-up of an electrical potential, discharged in lightning (see p. 64) and thunder. An hour or so before the storm, the barometer usually shows a very rapid fall.

Causes. Storms are the result either of a very quick drive upwards of warm air to great heights (thermal storms) or of the coming together of a warm moist air mass with a well-developed cold front (frontal storm). Strong air movement takes place with winds reaching to altitudes of 10,000 metres (32,800ft). The cooling of the elevated air masses leads to heavy condensation. The storm clouds reach heights at which temperatures are below 0°C, as shown by precipitation in the form of ice crystals or super-cooled droplets. Usually both forms are present and the alternation between ice and water droplets produces a high electrical potential. The negative and positive charges are separated, owing to the different speeds at which ice and water fall. The clouds act as generators. When the air becomes overcharged, the pressure is released as lightning and a new potential is generated. The sudden increase in pressure in the path followed by and heated by the lightning leads to a shock wave heard as thunder. The speed with which the sound of thunder spreads at 330 metres (1083ft) per second makes it possible to gauge the distance from the observation point of the storm.

Effects on weather. Thermal storms are local events resulting from intensive radiation. The rise of the warm moist air is associated with a strong wind lift, particularly over mountain slopes. Thermal storms, usually summer events, often last only for an hour or two before the sun breaks through again. Frontal storms occur when, in association with westerly conditions, a cold front penetrates and meets a warm air mass. The warm air was previously heated and has absorbed moisture while in a high pressure area. The cold front pushes the warm air before it, accentuating the sultry weather. Turbulence then develops strongly along a front often a few hundred kilometres long, moving eastwards. Frontal storms are more frequent on the coast than inland, and are not limited to summer. Indeed, in winter they are associated with blizzards.

Forecast. A local thunderstorm should not involve a permanent deterioration in the weather. It brings a fall in temperature but that does not last long. If there is no shift in the wind, a further period of stability can be expected. With rapid evaporation of the rain, however, storm conditions can be re-established quite quickly – even on the same day. If the fall of temperature is very marked, the rain will then be more persistent, normally associated with a wind shift from east to west. Frontal storms bring on a marked change in the weather, establishing westerly conditions.

Towering cumulus clouds heralding a storm (*above*)

View through a storm front (*below*)

Stratus clouds

Recognition. Stratus clouds can probably most frequently be seen when warm air approaches an area of lower-lying cold air – then the blue skies become paler, and a halo rings the sun (see p. 52). After a few hours, the veil of cloud becomes thick and grey. By the next day, the sun is no longer visible and the sky is obscured by a sheet of grey cloud. The cloud is either fibrous or ribbed, and has thin patches through which the sun shines occasionally, albeit very weakly. As the tendency to rain increases, the grey intensifies giving the impression that the clouds are coming ever closer to the ground. During and after the rain, cloud trails hang just above the ground. Ground fog and hill fog often occur with a stratus cloud sheet.

Causes. Stratus clouds form either with the rise of warm air over cold in association with a warm front, or as the consequence of radiative cooling – particularly with hill fog. The stratification is stable and horizontally uniform. The rising air along the front becomes colder and more moist, and on clear winter nights the strong radiation also leads to cooling. In both cases, condensation thus takes place and clouds are formed.

Stratus clouds occur in three levels of the troposphere:
1. High stratus cloud: cirrostratus (Cs), at heights of over 6000 metres (19,700ft). This appears as a soft white veil, usually fibrous. Such clouds consist of ice crystals and form haloes.
2. Middle height stratus clouds: altostratus (As), at heights between 2000 and 6000 metres (6600 to 19,700ft). These clouds have a grey or bluish colour, and appear structureless or ribbed. No halo phenomena occur.
3. Low stratus clouds: nimbostratus (Ns), at heights below 2000 metres (6600ft). A dark-grey rain cloud bringing rain in summer, snow in winter. A low and continuous cloud in association with ground fog or hill-fog is also called stratus (St).

Effects on weather. Stratus clouds are particularly associated with westerly weather conditions. The sequence cirrostratus, altostratus and nimbostratus characterizes the front of a depression (see p. 42) and the deterioration in weather. The rainfall produced by the gently-rising warm air is heavy and, at low temperatures, turns to snow. At the front, as the cold air retreats, frontal fog occurs. Winds are moderate, veering from southeast to south and southwest.

Forecast. The high stratus clouds are of particular interest. Before the arrival of the cirrostratus sheet, the observer will notice wisps of cirrus, the early portents of a disturbed weather situation. The build-up of cirrostratus, the lowering of the cloud base and the transition to altostratus cloud are certain indicators of an inbreak of bad weather by, at the latest, next day.

Altostratus cloud, together with altocumuli

Sunshine

Recognition. The sun, the central star of our solar system, radiates heat and light. The intensity of the energy received from it is dependent on the cloud cover, the height of the sun over the horizon and the season of the year. Intensity also increases with altitude, although without any perceptible extra effect, so that on mountains the ultra-violet radiation can cause sunburn. As radiation passes through the atmosphere its intensity is altered. Pollution and water vapour in the atmosphere produce the changes which result in the red colours of dawn and sunset.

Causes. The sun is a fixed star (see p. 16). If all its rays reached the earth without being absorbed, sunlight would appear white. It combines a mixture of the spectral colours from red to violet with wavelengths of 0·8 to 0·4 micrometres (μm – thousandths of a millimetre). The central point of the spectrum is near green-yellow (0·55μm). In addition to visible light, the solar radiation has a long-wave component in the infra-red range, not visible to man but amounting to about half the radiant energy, and a short-wave component consisting of ultra-violet and Röntgen radiation. A considerable part of the short-wave radiation is absorbed in the upper atmosphere (see p. 15). The mean value for the radiant energy received in the upper level of the atmosphere is known as the solar constant:

$$1·9 \text{ cal/cm}^2 \cdot \text{min} = 1·37\text{kW/m}^2 = 1·37 \cdot 10^6 \text{ erg/cm}^2 \cdot \text{sec}$$

Because of the elliptical orbit of the earth around the sun, the radiant energy received varies by about 3·5 per cent of the average value.

Effects on weather. The sun provides all the necessary energy for weather events to take place in the atmosphere. It is thought that physical-chemical processes on the sun affect the radiation intensity and therefore weather patterns. The more or less oblique impact of the sun's rays on the earth results in the various climatic zones of the earth, as well as the seasons. The greatest amount of energy is conveyed when the sun is vertical or nearly vertical. Twice a year the sun is visible at its zenith in the tropics. In the temperate regions, the sun always appears at an angle – in summer the angle is more acute, in winter the angle is greater. In polar areas the sun does not rise in the winter, whereas in summer it does not set. The angle of the sun is, however, very small in the polar regions so that, despite the long day, little energy is received. In mountainous terrain, the radiation intensity also varies seasonally, for in winter the mountains cast the valleys in shadow, in general making the winter longer and harsher.

The duration of sunshine is also influenced by the amount of cloud. In tropical arid regions, such as Egypt and Arizona, the average relative sunshine duration is 80–90 per cent. In central Europe, it is 40–50 per cent from May to August, but only 10–20 per cent from November to January.

Forecast. Floating particles in the air (water vapour and dust) decrease the strength of radiation reaching the earth's surface. The constant movement of the atmosphere is an important factor in the various weather processes (see p. 54). In long range forecasting, sunspot statistics are used, but an acceptable explanation together with an assessment of the effects of sunspots has yet to be found.

North Sea Canal at Ijlmuiden, Netherlands

Waves, motion of

Recognition. On inland waters high waves can be whipped up during storms, but waves more than a metre (3ft) high generally occur only at sea. On the ocean's surface, paradoxically, one can see that, although an increase in the wind's strength makes the wave length longer, the waves actually lose height (amplitude), particularly when there is no coast to break them. The longest waves in the southern hemisphere between Tierra del Fuego and New Zealand are about 800 metres (2625ft) but their amplitude only 20 metres (65ft). Even when no wind is blowing, waves can be seen which originate from some distant wind centre. The shallower the water, the closer the wave crests come together, giving short steep waves which break before the coasts. This is how the big 'breakers' are formed.

Causes. Waves are caused by wind. Those waves resulting from winds still blowing in the area are known as wind-waves, those from winds elsewhere as swell. The motion of the sea is composed of a number of waves together, differing in wavelength, height, period and direction. Wind-waves are determined by the duration of the wind, and the fetch (distance over which sea and wind are in contact). Cross winds cause the most turbulent waves.

Motion of sea at various wind strengths

	windstrength 2 light breeze	windstrength 6 strong wind	windstrength 10 severe storm
fetch	0·56 sea miles	140 sea miles	1570 sea miles
wind duration	0·7 hours	15 hours	73 hours
wave period	1·4 seconds	7 seconds	14·7 seconds
wave length	2 metres (6·6ft)	51 metres (167·3ft)	225 metres (738·2ft)
wave height (amplitude)	0·05 metres (0·16ft)	2.5 metres (8·2ft)	15·8 metres (51·8ft)

If the depth of water is less than half the wave length as, for example, near a coast, the waves become surf or breakers. The wave characteristics are the result not only of the pressure and force of the wind but also of differences in atmospheric pressure and of friction. The latter also affects the ripples due to icebergs.

Effects on weather. The strength of the waves is related to the weather situation and the associated winds. For inland waters it is mainly a question of storms, but weather situations with fall or Föhn winds and cold air inbreaks (northerly weather conditions) are capable of producing strong waves. This is also true of the westerly weather situation which produces gusty weather even well inland, strong winds on the coast and winds of storm or hurricane intensity (see Beaufort Scale and windspeed, p. 112) over open seas.

Forecast. The state of the sea's surface provides a method of determining windspeed: gentle ripples running up a beach on a summer's day are indicative of a sea breeze (see p. 32) characteristic of good weather conditions. A more disturbed sea with white crests shows a squally wind.

Freighter in a swell, but with a gentle wind (*above*)

Wind-waves in a storm (*below*)

Wind

Recognition. A large number of different air movements can be observed. In the mountains during the day a breeze develops towards the mountain and, comparably, on the coast a breeze blows from the sea to the shore. Then there is the squall heralding the arrival of cold air or a storm. Another form of air movement is the fall wind, the best-known example of which is the Föhn (see p. 46). In deep depressions, storm or hurricane winds develop with great destructive powers. Over the open sea, in particular, the wind blows with unrestricted force. There are also periodic winds with a seasonal response, for example, the Indian monsoon in summer and winter. The trade winds are another wind system of large dimensions and in the tropics, particularly at sea, they blow with great regularity and little interruption.

Causes. The fact that the wind blows from places with higher pressure to places with lower pressure in order to equalize the pressure is sufficient explanation only for very local air movements. The wind does not blow only in the horizontal plane – there are vertical winds, too, such as those occurring during a storm. In addition to differences in temperature that give rise to differences in pressure, there are other forces associated with air movement. These forces can be traced back to three causes:

1. The earth's rotation
2. Air movement in a circular route
3. Friction with the earth's surface

The mechanics of land and sea breezes is caused by temperature and pressure differences with a daily rhythm. Large air movements show the effect of the Coriolis force – the displacement of air streams by the turning of the earth. Such a large wind system is that of the trade winds, north and south of the equator. In addition to rotation, the differing energy provision of air masses and the distribution of land and sea are important. Windspeed is measured in metres per second or knots (sea miles per hour). See pp. 112–20.

Effects on weather. Apart from local storms, the strongest winds are experienced in westerly and northerly weather systems, with the arrival of cold air. The windspeeds increase with height. The stronger the wind the more squally it is, dependent on the ground conditions and the stratification in the atmosphere. The strongest squalls come with cold air inbreaks. Over the sea the winds are more regular, since the irregularity of the ground conditions is missing.

Forecast. The Dutch meteorologist, Buys-Ballot, proposed the following law 'If I stand in line with the wind direction, the wind at my back, to the left and in front of me is a depression with predominantly bad weather, to the right and behind me is a high pressure area and better weather.' In summer westerly and northerly winds bring moisture and cool conditions, southerly and easterly bring dryness and warmth. A sudden change in the wind means a change in the weather. If the clouds are at different heights and move in various directions, this is evidence of different winds at various heights. Usually the upper wind determines the weather sequence. A freshening wind in the evening, particularly from southwest or northwest, is an indicator of a tendency to rain; in winter, after a period of frost, it is an indicator of a thaw.

Beginnings of a water spout, extending from a large cumulus cloud (*above*)

Dust storm off western Africa – Harmattan wind (*below*)

Stevenson screens (see p. 101) and measuring instruments at a meteorological station

Basic meteorology

An outline of the forces that contribute to the weather has already been given in the section, *What is Weather?*, p. 13. For weather observation and forecasting, the most important meteorological features are air pressure, air temperature, air humidity and air movement. These are the most easily observed and understood by the non-professional. A large number, although not all, weather events can be described and explained with the assistance of these features. These subjects need to be explained in greater detail, and, since their measurement and recording are of vital importance, we shall first describe the practical aspects of the apparatus used.

Air pressure

The most commonly used apparatus for measuring air pressure is the aneroid barometer. This consists of a metal container (made of steel or beryllium) from which the air has been evacuated. When the air pressure increases, this box is compressed and, when the pressure falls, the box recovers elastically. The compression and extension are transferred to a recording pointer by a lever mechanism. The significant weather features to observe are:

1. The rise or fall in air pressure
2. The rate of change

The barometer is therefore marked with a scale, on which the position of the pointer can be read and changes noted. The added indications on many domestic barometers of 'dry', 'fair', 'changeable', and 'rain' are of little significance for meteorologists. When changes in the air pressure are to be related to a regular record-keeping system, then the standardized scale must be used.

either
millimetres of a mercury column (mm), inches (ins),
or millibars (mb)

as already discussed on pp.14–16. A good mercury barometer is preferable for the standardized observations. In English-speaking countries, the measurements are still often given in inches. The table on p. 96 shows the relationship:

mm mercury	millibars	ins mercury
720	960	28·35
724	965	28·5
727·5	970	28·65
731	975	28·8
735	980	28·9
739	985	29·1
742·5	990	29·25
746	995	29·4
750	1000	29·5
754	1005	29·7
757·5	1010	29·8
761	1015	30
765	1020	30·1
769	1025	30·3
772·5	1030	30·4
776	1035	30·55
780	1040	30·7

In practice, recorded air pressure can be converted from mm to mb and vice versa by using the following ratio:

$$\begin{matrix} 1mm = 1\cdot333mb \\ 1mb = 0\cdot750mm \end{matrix} \quad or \quad \begin{matrix} 1in = 33\cdot8667mb \\ 1mb = 0\cdot02953in \end{matrix}$$

A simple means of registering air pressure changes over a long period is to mark the recordings on a graph with days on the horizontal axis, say, 5mm (0·2in) per day, and air pressure on the vertical. The base line should represent 760mm or 29·9in (= 1013mb). The barometer reading should be taken at the same time every day. A sequence of pressure changes at that position over weeks and months can then be recorded.

The attentive observer will notice that high pressure is not always associated with good weather.

Graph paper recordings can also be made mechanically by using an instrument known as a barograph. In order to accommodate sufficiently large movements, the barograph is made up of several metal containers. The movement is transferred to a pen by levers, the pen being filled with a special ink to record on the special graph paper. The paper is mounted on a slowly-turning drum driven by clockwork (one revolution per week). The barograph should be located in a peaceful undisturbed place, to ensure that its recordings are accurate.

Another practical piece of apparatus that registers changes in air pressure is the altimeter, often carried by mountaineers. With an altimeter at hand, the user can instantly appreciate the fall in air pressure resulting, as we have seen, from increased height. Some altimeters, in fact, have a dual scale which is marked in both mm (in) or mb, as well as in height. This makes it possible to establish the absolute, average and reduced pressure (i.e., reduced to sea level). By noting the position of the altimeter at any given moment, short-phase changes in pressure can be detected. The use of such a piece of apparatus makes it possible to observe the barometric tendency, that is to say, whether pressure is rising or falling and how quickly. It must be remembered that the altimeter will also be registering change in height, and a good altimeter registers to plus or minus 10 metres. The change of pressure with height has no bearing on weather events.

Absolute pressure. The pressure present at the place in which the recording is made. For weather forecasting, comparison with measurements at other stations is important (see reduced pressure).

Average pressure. The pressure expected at a given point, based on the fall of pressure with height in a normal atmosphere. At 0 metres (0ft) altitude 760mm (29·95in) = 1013mb; at 500 metres (1640ft), 716mm (28·2in) = 954·5mb; at 1000 metres (3280ft), 674mm (26·55in) = 899mb; etc. In the lower atmosphere the reduction is 1mb per 8 metres (26·25ft) rise.

In the lower layers of the atmosphere, air pressure falls 1mm every 11 metres ($\frac{1}{10}$ in every 90ft). Climbing a mountain with a barometer at hand shows this quite dramatically, particularly with an altimeter – an aneroid barometer with a height scale rather than a pressure scale. The altimeter also shows weather-induced changes in pressure, and there are altimeters with the necessary calibration for both purposes. Since the pressure varies from place to place and from time to time, an altimeter must be standardized at a place of known height in the area before being used.

Reduced pressure. The pressure recording at sea level corresponding to the recording taken at the station at greater altitude, used as a means of comparison. The average pressure at sea level is 760mm (29·95in) = 1013mb. It is best to calibrate a barometer when air pressure, according to local weather stations, is relatively constant. For example, a weather report might include an announcement that 'air pressure reduced to sea level 765mm (30·1in) = 1020mb'. This recording should be used to correct the reading by means of the adjustment screw on the back of most barometers. It can also be calculated from the height of the station.

Air pressure shows a daily rhythm, rising in the early morning and in the evening, and falling from midnight until the early morning and from late morning until late afternoon, the last being particularly marked in summer because of the warmth. These daily changes, amounting normally to 1 millibar, are of no significance to the weather. They are more similar to the changes in tides, explicable in terms of the position of the moon and sun. They are more in evidence during periods of relatively constant weather. In mountainous country, the opposite tendency is more often revealed – with a rising pressure in the afternoon as the air rises from the valleys.

In the monthly averages, it is more normal to register higher pressures in August and September and a lower pressure in April. In the mountains higher pressure comes in summer, lower in winter. These changes are associated with weather events – high pressure systems are more frequent in the autumn, but low pressure systems tend to predominate in spring. Winter depressions in the mountains are significant.

The barometer registers the weight of the air column at the place of observation. Cold air is heavier and is therefore recorded as a higher pressure. Warm air is light, the barometer therefore registers falling pressure. This ratio of pressure change is also important in weather forecasting.

barometer	millibar per hour	weather
rising	0·25–0·5	high becoming established long period
rising	1–2	wedge, short period
falling	0·25–0·5	low approaching, long period
falling	1–2	stormy situation in summer thunderstorms

There are many exceptions from the principles outlined in the table. Two examples are as follows:

The barometer shows a *falling tendency*, but the weather remains fine. Warm air from the south makes its way northwards at ground level. The barometer reacts to this by falling. The warm air near the ground moves forward particularly quickly (quicker than at altitude) and, at the rear, cold air coming from above acts as a replacement. The sinking cold air becomes cloudless, and the sun shines through.

The barometer shows a *rising tendency*, but the weather is cloudy and rainy. Here it is a matter of cold northerly air moving quickly southwards. This is registered on the barometer as rising air pressure. The cold air lifts the warm air in its advance, but this entails cooling. The recognized consequence is cloud and rain.

After the passage of a depression, the wedge of cold air brings about a *rapid rise in air pressure*. Despite the rise there are storms, rain or snow. Here, also, the warm air is lifted and a cooling effect ensues.

Finally, during a long persistent high pressure, a further *rise of pressure* is witnessed. At the same time, it becomes less sunny and visibility is decreased as a haze forms (see p. 54).

Air pressure is shown on maps by isobars, lines joining all places with the same pressure. Usually on weather maps, they are drawn at every 5mb. Obviously, values recorded for stations at various altitudes are reduced to sea level before mapping.

A high pressure area is shown by an encircling series of lower value isobars. The depression is shown by higher value isobars on all sides. The distance between isobars is of particular interest, since this shows the pressure gradient.

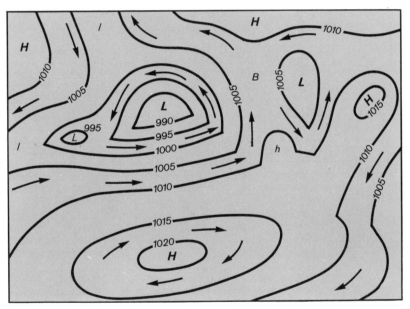

Part of a weather map with isobars showing a high and a low. Isobars are every 5mb. The arrows show wind direction: the air movement is clockwise around the high, anticlockwise around the low. The cause of this is the Coriolis force (see p. 118). This effect of the earth's rotation favours the longer persistence of pressure systems.

H = high   L = low   h = high pressure ridge   I = extension of low pressure

distance between isobars	pressure difference	weather
small	great	turbulent
large	small	calm

The gradient is measured as the rise/fall in pressure (mb) over sixty sea miles at right angles to the isobars. An example is the rise from 1005mb to 1010 over fifty sea miles. The calculation is:

$$50:60 = 5:X \qquad X = \frac{5 \times 60}{50} = 6$$

The gradient is 6mb in 60 sea miles.

For large air movements, the rotation of the earth is important. The movement associated with given air pressure distributions can only be understood by reference to the so-called 'Coriolis force' (see p. 118).

Air temperature

Most thermometers in general use contain mercury or alcohol – or are bimetallic. The mercury or alcohol thermometers depend on the expansion and contraction of mercury or alcohol resulting from heating and cooling. The bimetallic thermometer consists of two metal strips, welded together and curved, the two metals having different co-efficients (constant factors) of expansions. They are often adjustable and are used mainly as thermographs, continuously recording the temperature, in a similar manner to that in which the barograph records pressure (see p. 96).

For weather observation, the maximum and minimum thermometer is particularly useful, recording the highest and lowest temperatures during the observation period by means of a metal indicator moved by the mercury or alcohol.

The thermometer is graduated from the freezing point to the boiling point of water – the length of scale between is divided evenly. In the celsius (or centigrade) scale, named after the Swedish astronomer Anders Celsius (1701–1744), the scale between freezing point and boiling is divided into 100. On the fahrenheit scale, named after Danzig-born physicist Daniel Gabriel Fahrenheit (1686–1736), there are 180 divisions, and on the réaumur scale, named after the French biologist and technologist René-Antoine Réaumur (1683–1757), there are 80 divisions. The fahrenheit scale is still used, to some extent, in Britain and other English-speaking countries, but the centigrade scale is becoming more and more international because of its use of the decimal system.

Comparative table of thermometer scales

	celsius	fahrenheit	réaumur
	−25°	−13°	−20°
	−15°	+ 5°	−12°
	− 5°	+23°	− 4°
	0°	+32°	0°
	+ 5°	+ 41°	+ 4°
	+10°	+50°	+ 8°
	+15°	+59°	+12°
	+20°	+68°	+16°
	+25°	+77°	+20°

Fixed points on the thermometer scales

	celsius	fahrenheit	réaumur
freezing point	0°	+32°	0°
boiling point	+100°	+212°	+80°

To check the accuracy of a thermometer, place it as deeply as possible in a container full of ice. Since zero point corresponds to the temperature of melting ice, the thermometer should read zero. It should then be placed in warm water at about 35°C and compared with the reading of a medical thermometer placed in the same water. Since the medical thermometer is very carefully calibrated, it can be used in this way as the standard.

When measuring air temperature, accuracy depends not only on the reliability of the instrument but also on its location. The thermometer should be in a well-ventilated

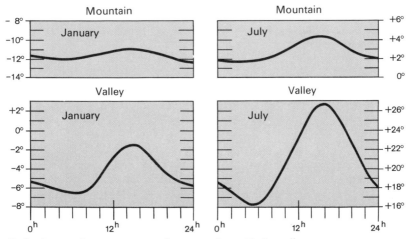

Daily change of temperature on the mountain and in the valley

place and not in the sun. The site should be away from houses. Usually meteorologists place their thermometers, together with other instruments, in a well ventilated, white painted box or container – opening to the north. This box is known as a Stevenson screen, is set 3½ft above grass in Britain. In central Europe, it is set 2 metres (6½ft) above the ground (see p. 94).

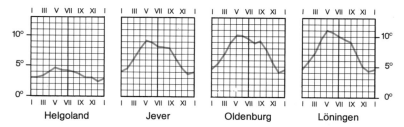

Helgoland   Jever   Oldenburg   Löningen

Average daily temperature change on the coast and inland. The range increases with distance from the coast as the moderating influence of the sea is lost. Helgoland is an island, Jever, Oldenburg and Löningen are respectively 11 km, 30 km and 80 km from the North Sea coast of West Germany.

The daily range of temperatures is greater over the land than over the sea. In both cases the minimum occurs shortly before dawn, the maximum at about 14.00 hours. With undisturbed incoming and outgoing radiation, the range over land (maximum and minimum) can be 5°C to 10°C, but over the open ocean the range is generally only 1°C. This is caused by the different warming characteristics of land and sea (see p. 168).

The temperature range during an entire year will also show marked differences between land and sea.

place	a mean January temperature	a mean July temperature
Yakutsk, Siberia	−42·9°C	18·8°C
Thorshaven, the Faroes	3·2°C	10·8°C

Within Europe, the east has cold winters and hot summers under the influence of a continental climate while the west, particularly the British Isles, has cool summers and mild winters caused by the Gulf Stream and a maritime climate.

Temperature is also lowered as altitude increases, otherwise the presence of snow on high mountains in tropical areas, such as Kilimanjaro 5895 metres (19,340ft) in East Africa, would be difficult to explain.

Although warm air often penetrates the upper layers of the atmosphere, the temperatures in the lower and middle troposphere decrease by 0·65°C per hundred metres

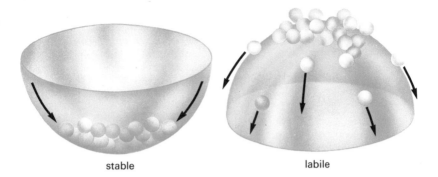

stable   labile

A stable system is one in which a position is maintained (*left*, the balls in the bowl roll together). A labile system is one with a tendency to change (*right*, the balls on the bowl will roll in various directions apart from each other). Stable and labile stratifications are very significant in weather. A labile or unstable stratification, for example, may occur when higher levels of cold air move over layers of warm air. An example of stable stratification could be caused by cold air layers at ground level. Various temperate air masses can become stable or unstable very quickly. A closed structureless sheet of cloud is indicative of stable conditions, while towering cloud masses with interspersed patches of blue sky are indicative of instability (storms, rain showers).

(328ft) up to about 10 kilometres (6·2mi). In the upper troposphere and in the stratosphere (see p. 131), the temperature remains constant or rises once more. If the temperature at any level begins to increase, then meteorologists refer to this as an inversion (see p. 62). With ground inversions, the temperature rises from the ground up to a definite height – and then diminishes. Ground inversions begin in the night when, with suitable weather conditions, out-radiation takes place. An inversion at altitude is associated with a decrease in temperature from ground level to a definite height – and then increases (lower level of the inversion). The increase reaches up into the atmosphere and is terminated at the upper limit of the inversion. From this upper limit, there is then a further fall in temperature as the altitude continues to increase. For weather events, the differential heating of neighbouring air masses is particularly important. This means that air temperature differences over land and over sea lead to the development of differences in pressure (see p. 32). Movement of the air masses generates the winds. Not only are pressure and temperature interrelated but so also are air temperature and water vapour. This is because of the moisture content of air and because of the ability of air to hold a quantity of water vapour specific to a given temperature (see p. 106).

An analysis of weather based simply on the measurement of temperature is too limited. In summer, with a fall in temperature, rain can be expected, but in winter a fall probably is indicative of good weather. The significance of rising temperatures is also dependent upon the season. In general, a rapid departure from the daily temperature sequence indicates a change in the weather, as, for example, with a sudden rise in temperature in the morning while warm air is arriving. The moisture content of the warm air is also significant.

In a similar way to that in which air pressure is mapped (see p. 99), places with the same temperature are linked by isotherms. A look at these shows the importance of the distribution of land and sea, as well as the anticipated effect of the latitude.

The earth's surface, warmed by the sun, heats the layers of the atmosphere near the surface by outward radiation. These layers become labile, and vertical movements of warm air take place. The degree of warming of the lower layers depends on the nature of the surface (fields, settlements, lakes, forests, etc.) and on the creation of thermal currents.

The form of the land influences the development of thermal currents, as, for example, when heated air from the lower layers impinges on a woodland. The moist air is lifted and cooled, forming cumulus clouds.

On mountain slopes, sunshine warms the air and causes an up-slope wind. The movement is strong and is followed by the formation of cumulus cloud. Such rising air, known as thermal, is extremely important for aircraft, hang-gliders and birds, especially those which hover and glide, and kites.

Air humidity

Meteorologists make use of two principles in measuring humidity:

Hair hygrometer. This instrument employs a characteristic of human hair and of some artificial fibres – their tendency to stretch in moist air and contract in dry air. This movement is transferred to an indicator. The hair hygrometer is covered with a damp cloth for a half hour before being used and then adjusted to 95 per cent relative humidity. In a thick mist an accurately-calibrated hair hygrometer should read 100 per cent.

Psychrometer or **wet and dry bulb thermometer.** This instrument makes use of the fact that there is an increase in evaporation when dry air is present, and a reduction when the air is moist. The apparatus consists of two identical thermometers, the bulb of one wrapped in cotton with one end of the wrapper immersed in water to keep it saturated – the wet bulb. The other thermometer bulb is not wrapped. The air around the wet bulb causes evaporation, and the evaporation requires warmth. The 'wet' thermometer therefore shows a lower temperature than the 'dry', and from the temperature difference the humidity can be calculated. The difference falls to 0° when the relative humidity is 100°. The relative humidity is read off by using a table. An explanation of the distinction between absolute and relative humidity is given in the section on Humidity (see p. 58).

Both pieces of apparatus measure the air's intake of water vapour, but this is only apparent to the eye when the saturation point is reached and condensation takes place. Mist, clouds and rain are the most visible manifestations of water vapour in the atmosphere. Temperature contributes to this process (see p. 76). Cooling of the air leads to condensation, since air at a given temperature can hold only a certain amount of water vapour, as in the following example:

Moist warm air rises. Its temperature is 25°C and its water vapour content is 15g/m³ (0.015oz/ft³). This corresponds to a relative humidity of 65 percent (see table and formula p. 58). As the air cools to 20°C, the relative humidity rises to 87 percent. The formation of clouds will begin at an altitude at which the temperature falls to between 17°C and 18°C (condensation level, dewpoint level). Rising air cools at 1° per 100 metres (328ft) so long as no liquid water is formed (unsaturated air). In this example the clouds will form at about 700–800 metres (2300–2625ft).

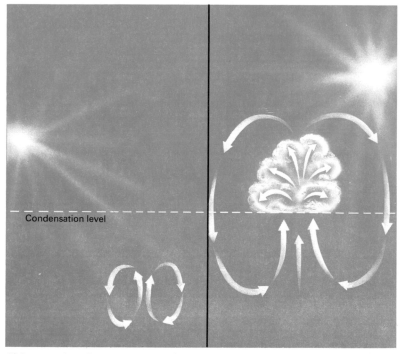

Condensation level

If the warming of the air near the ground is restricted, the rising air currents do not reach condensation level. If the rising air reaches a level at which cumulus cloud is formed this is the condensation level (*right*). Decisive in this are temperature and relative humidity. The autoconvective temperature is that necessary for air to rise automatically from the ground until condensation takes place.

Friction and the humidity of warm air currents depend upon the nature of the earth's surface. Various condensation levels therefore result, and cumulus forms at different heights. These different thermal levels are noticeable particularly in mountainous regions, such as the Alps.

In sunny weather the air near the ground will become progressively warmer during the day. It will thus become gradually lighter than the cold air above, and rise. If the conditions are such that the air rises automatically to the condensation level, then the ground layers have reached what is called the 'autoconvective' temperature. The determination of this temperature depends on an exact knowledge of the temperature and water vapour conditions at higher levels, and is essential for aviation and weather forecasting.

The importance of pressure and temperature to humidity is shown by the variations which take place during the course of the day, even when there is no change in the existing air mass. The hygrometer will show a low relative humidity on a warm summer's day but, by evening and in the night, the humidity will rise. Often it will reach 100 percent and dew will form. The presence of clouds can hinder this fall in temperature and rise in relative humidity by blocking outward radiation.

The dewpoint marks the expected lowest temperature of the night, given of course that there is no change during the night in the composition of the air mass resulting, for example, from the arrival of a cold air stream. If the dewpoint is near 0°C, then a night frost occurs. High relative humidity on a starlit night often brings fog. If the air is dry, then lower temperatures can be expected.

Observation of low temperatures near the ground shows that frosts occur at ground level, whereas temperatures 1 to 2 metres (3.3 to 6.6ft) above ground level are still a few degrees above zero. The danger of frost is greatest in hollows. Establishing the dew-point is important in forecasting frost. The necessary temperature and moisture measurements must be made methodically after sunset. In order to understand these processes, a knowledge of the relationship between air humidity and air temperature is necessary.

temperature	water vapour content	saturation moisture level	air humidity
rising	constant	rising	air relatively dry
falling	constant	falling	air relatively moist

Air density

Before the development of rockets and satellites, investigation of the composition and density of the atmosphere at high altitudes was carried out only by means of hydrogen-filled balloons and radio sondes (small radio transmitters). The latter are connected to appropriate measuring instruments, which send out a signal that varies according to the weather element at each level and that can be interpreted to provide temperature, pressure and humidity readings. The track taken by the radio sonde indicates the speed and direction of the wind. At a height of about 16km (10mi), the balloon carrying the sonde bursts and the equipment floats down to earth on a parachute. The laws of physics lead us to conclude that the density of air diminishes with altitude. There must also be variation in its composition, since the gases of which the atmosphere is composed have different weights (see p. 11). Light gases, helium and hydrogen, for example, must be present in relatively greater proportions at high altitudes and, conversely, the heavier gases, nitrogen and oxygen, must predominate in the lower layers.

The pressure of the atmosphere is an important factor in the movement of air. Air density is proportional to pressure, and inversely proportional to temperature. In hot regions or those with high summer temperatures, as well as in high altitude airports, low pressures are recorded. Aircraft then require a longer take-off or cannot be fully loaded.

Calculation of air density:
$$\text{air density in kg/m}^3 = \frac{0.349 \text{ air pressure in mb}}{273 \pm \text{ temperature in }^\circ\text{C}}$$

Today we know that up to a height of almost 100km the composition of the air varies very little. This is explained in terms of a large zone of mixing in which the exchange of

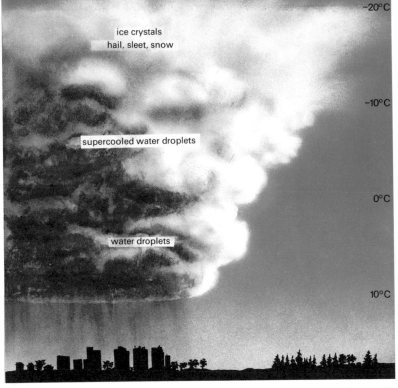

-20°C

ice crystals
hail, sleet, snow

-10°C

supercooled water droplets

0°C

water droplets

10°C

In summer the contrast between warm and cold air is particularly marked, and towering cumulus clouds are formed with rising currents from 6000 to 10,000 metres (19,700 to 33,000ft). The rapidly-rising warm air produces storm clouds (cumulonimbus). The speed of the rising air (thermals) can reach 10 metres (32.8ft) or more per second. In the lower third of its total expanse, the storm cloud is made up of water droplets, in the middle section, supercooled water droplets and, in the upper third, ice crystals.

gases takes place and thus separation according to specific weight is prevented. As observed, a reduction of air density with altitude is comparable with the reduction in pressure. At sea level, the density of air is 1.29 grams/litre. At 17km (10.6mi) altitude, the density is about one tenth of this. The decrease of density for above 100km (62mi) is similar to the decrease in pressure but more gradual.

Air density is subject to diurnal and seasonal cycles of change, and there are further changes reflecting the activity of the sun.

Near the ground the addition of dust, the products of combustion and other byproducts have little effect on density, but such additions to the atmosphere are important in bringing on several weather types, such as rain, mist and fog.

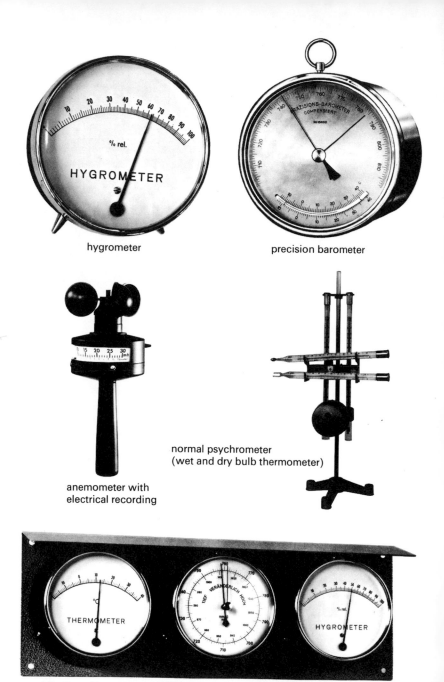

hygrometer

precision barometer

anemometer with
electrical recording

normal psychrometer
(wet and dry bulb thermometer)

combination of instruments: thermometer, barometer, hygrometer

combined hygrograph, thermograph and barograph

sunshine recorder

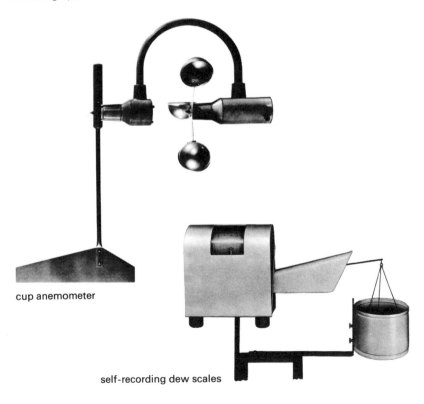

cup anemometer

self-recording dew scales

Wind measurement

A wet finger can indicate the direction of the wind – the windward side dries first – but simple equipment in the form of a windsock or wind cone is more reliable. This consists of a hollow cone made from a textile material. The larger end is fastened to an open vertical rigid ring. The smaller end is also left open to prevent the sock from fluttering. The rigid end of the wind sock is fixed to a vertical pole on a roof or on open ground. The wind direction is the direction from which the wind blows, the sock trailing in line with it to the opposite quarter of the compass.

The windvane or weathercock has been used as an indicator of wind-direction for a long time, and there are now electrical and mechanical methods of registering the direction readings.

For the measurement of windspeed, various methods and devices are available. One of the best known is the cup anemometer, which can be used in the hand. The principle involved is the movement, in a clockwise direction, of four cups of metal or plastic, mounted on arms. The stronger the wind, the more quickly the cups rotate. If the number of revolutions is counted, a measurement of windspeed is obtained. Manufactured anemometers are supplied with a scale which gives speed in metres (ft) per second. The range of hand-held cup anemometers is generally 0–25 metres (0–82ft) per second. This corresponds to 0–9 on the Beaufort Scale (calm to strong gale).

Beaufort Scale no	Beaufort Scale and windspeed character of wind	effects of wind inland
0	calm	calm, smoke rises vertically
1	light air	wind direction shown by smoke but not by flags
2	slight breeze	wind felt on face; leaves whisper flags move
3	gentle breeze	leaves and thin twigs move; pennants extended fully
4	moderate breeze	paper and dust lifted twigs and thin branches move
5	fresh breeze	small trees in leaf begin to sway, white horses on lakes
6	strong breeze	thicker branches move, whistling in telephone lines, umbrellas difficult to use
7	moderate gale	whole trees moving; resistance to movement against wind perceptible
8	gale	twigs broken from trees; movement in open difficult
9	strong gale	minor damage to houses (chimney pots and tiles)
10	whole gale	trees uprooted, major damage to houses
11	storm	widespread damage (rare)
12	hurricane	severe destruction

Wind direction is indicated by the direction from which the wind blows – the west wind blows from the west, towards the east. The wind rose shows the wind directions in degrees from north by east to south and west. In meteorology, the direction is usually given in degrees as part of the full 360° of the circle (see p. 115). At sea older divisions are still employed.

International table, accepted as from 1.1.1949

Beaufort Scale no	effect of wind on the sea	pressure in kg/m^2
0	sea like a mirror	0
1	scale-like ripples without foam crests	0–0.1
2	small wavelets, still short but more pronounced crests that do not break	0.2–0.6
3	crests begin to break, foam of glassy appearance, perhaps scattered white horses	0.7–1.8
4	small waves, becoming longer, fairly frequent white horses	1.9–3.9
5	moderate waves with more pronounced long form, many white horses; small chance of some spray	4.0–7.2
6	large waves begin to form; crests break with patches of white foam; some spray	7.3–11.9
7	sea begins to tower; foam from breaking waves blown in streaks in wind direction	12.0–18.3
8	moderate high waves of considerable length; crests begin to break in spindrift; foam blown in well-marked streaks in direction of wind	18.4–26.8
9	mountainous waves, dense streaks of foam in wind direction; rollers begin; spray may affect visibility	26.9–37.3
10	very mountainous waves with long overturning crests; sea white with foam; heavy rollers exerting great pressures; visibility impaired by spray	37.4–50.5
11	exceptionally mountainous waves; much reduced visibility, caused by spray	50.6–66.5
12	air filled with foam and spray; sea completely white; severely reduced visibility; no distant objects discernible	66.6 and more

There are a number of electrical and mechanical devices for remote testing and recording of anemometers.

Air movement is described in terms of its direction and speed. As a rule the wind direction is given in degrees from north by east:

north wind = 0°
east wind   = 90°
south wind = 180°
west wind   = 270°

See also the wind rose on p. 113.

In sea navigation the point division is still used in part. A point is 1/32 of the compass circle, that is 11.25°. Each point has its own name. As an example, the names of the first 45° (the first half quadrant) are given below:

north   = 0°
north by east   = 11.25°
north-northeast   = 22.50°
northeast by north = 33.75°
northeast   = 45°

When windspeeds are quoted it can be in miles per hour, kilometres per hour, or sea miles per hour (knots) as well as in metres per second:

1 knot = 1.852 kilometres per hour = 0.515 metres per second.

The *Beaufort Scale* is quite widely used in indicating windspeeds (see p. 116).

Windspeeds of 12 on the Beaufort Scale (hurricane) are over 120km/hr (75mph). In fact, still higher speeds – over 200km/hr (125mph) – have been reached in typhoons, and these speeds were achieved at ground level. With increases in altitude, still higher speeds are reached. In the upper layers of the troposphere – at altitudes 6 to 15km (3.7 to 9.3mi) – jet streams have been identified at speeds of 400km/hr (250mph), which make flying extremely hazardous.

Air moves from a point of higher pressure to a point of lower pressure, and windspeeds are obviously related to these differences in pressure. The stronger the wind, the greater the pressure difference and the less the distance between the two points with the pressure differences. The greater the gradient (see p. 100), the greater is the windspeed. The daily sequence of weather is often responsible for the occurrence of pressure differences in temperate air masses. For example, air over the land warms more rapidly than air over the sea (see p. 168). Only in local movements are wind direction and windspeed determined by pressure differences. Over wide areas, all air movements resulting from pressure differences, following differential heating of the air masses, are subject to deflection owing to the rotation of the earth.

Upper and lower limits of windspeed and pressure in comparison to the Beaufort Scale

Beaufort scale	m/sec	km/hr	mph	knots	pressure in kg/m²
0	0– 0.2	1	1	1	0
1	0.3– 1.5	1– 5	1– 3	1– 3	0– 0.1
2	1.6– 3.3	6– 11	4– 7	4– 6	0.2– 0.6
3	3.4– 5.4	12– 19	8–12	7–10	0.7– 1.8
4	5.5– 7.9	20– 28	13–18	11–15	1.9– 3.9
5	8.0–10.7	29– 38	19–24	16–21	4.0– 7.2
6	10.8–13.8	39– 49	25–31	22–27	7.3–11.9
7	13.9–17.1	50– 61	32–38	28–33	12.0–18.3
8	17.2–20.7	62– 74	39–46	34–40	18.4–26.8
9	20.8–24.4	75– 88	47–54	41–47	26.9–37.3
10	24.5–28.4	89–102	55–63	48–55	37.4–50.5
11	28.5–32.6	103–117	64–72	56–63	50.6–66.5
12	>32.7	>118	>73	>64	>66.6

m/sec = metres per second
km/hr = kilometres per hour
mph   = miles per hour (1 mile = 1609 metres)
knots = sea miles per hour (1 sea mile = 1852 metres)

Pressure is that of the wind on a smooth vertical surface facing the wind. This method of recording is used in building regulations.

The comparative table above, giving windspeed and the Beaufort Scale, is based on measurement at the internationally-agreed height of 10 metres (32.8ft). The speeds relating to the Beaufort Scale will be about 20 percent less at, for example, 4m (13.1ft) above the ground, and about 20 percent more at 30m (98.4ft) above the ground. These standards have been in existence since 1949, before which time measurements were made at 6 metres (19.7ft) above ground level. Under hurricane conditions, individual

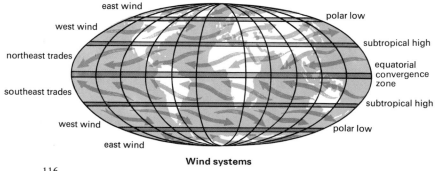

Wind systems

Vortices – powerful eddies of air – do not form solely as a consequence of the heating of the atmosphere near the ground. They can also be attributed to disturbance in an airflow when its passage is resisted by trees, houses, etc. With increasing horizontal speeds, the turbulence over such obstacles increases and leads to strong vertical disturbance.

measurements of windspeed are generally indicative of the force reached and cannot be used as a basis for calculation of speeds at other levels.

> In the northern hemisphere, all winds are deflected right
> In the southern hemisphere, all winds are deflected left

This deflection is greatest in the middle latitudes, notably over Europe, it is nil at the equator and increases with latitude. It also increases with windspeed.

The deflection is an example of the Coriolis acceleration, named after the French physicist C. G. de Coriolis (1792–1843), who described the principle: 'A moving body is deflected from its path if its movement is linked by inertia to the rotation of a revolving reference body'. The Coriolis force is a factor in the formation of the trade winds.

A further deflection and braking is suffered by wind near the ground caused by friction between the air and the earth's surface. Mountains dam back or deflect the movement of air. To sum up, windspeed depends on the following factors:

1. The air pressure gradient
2. Latitude (Coriolis force)
3. Loss caused by friction with the earth's surface
4. Curvature in the route of the wind

There is also a range of tropical winds and wind systems, associated with the Coriolis force (trade winds), the differential warming of land and sea (monsoon), and geographical features such as mountains (Föhn) (see also table p. 184).

Air movements are marked on weather maps by arrows moving in the wind's direction. In calm conditions, the circle on the map indicating the station is encircled with a further ring. Windspeed is symbolized by full and half feathered arrow shafts. A whole feather indicates two grades on the Beaufort Scale, a half feather one grade. Windspeeds of 10 or more on the Beaufort Scale are symbolized by a black solid triangle on the arrow.

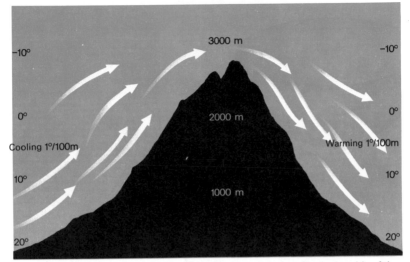

Winds blowing against a mountain will be lifted and cooled. On the lee side of the mountain the air is warmed as it descends, differing by 1°C for every 100 metres (320ft) of altitude both rising and descending. If the air is moist and condensation takes place, this rate is altered. Once the air reaches a relative humidity of 100 per cent, any further rise produces cooling at 0.5°C per 100 metres (320ft), since the heat is released by condensation. On descending, however, the rise in temperature is again at the rate of 1°C per 100 metres (320ft). If the air had lost moisture quickly in rising, then the descending wind will be a few degrees warmer in the valley (see Föhn winds, p. 46).

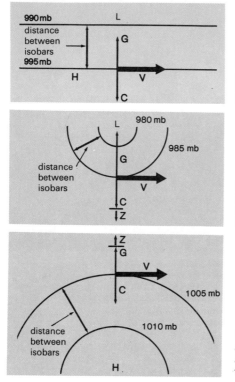

Air movement when a balance has been achieved between the gradient force and the Coriolis force (boundary zone between low and high)

Balance between gradient force and Coriolis force in a cyclonic system. Centrifugal force Z is the third force in operation here

Balance between gradient force and Coriolis force in an anticyclonic system. Centrifugal force Z is also present

Deflection force resulting from the rotation of the earth (Coriolis force)

A stronger wind blows in an anticyclonic system than in a cyclonic system, given the same distance between isobars. The deflection to the right of moving air by the Coriolis force in the northern hemisphere continues until a balance is achieved between the gradient force (rate of decline from the higher to the lower pressure) and the Coriolis force. This balance (see diagrams, below) is associated with a greater distance between isobars, as well as with a lesser curvature of the isobars in high pressure areas than in low pressure areas. This is true only for similar latitudes and windspeeds.

G = gradient force, C = Coriolis force, V = wind vector, L = low, H = high

All about weather maps

The weather forecaster's essential tool is the weather map. These are prepared according to an internationally-agreed scheme, employing a system of recognized symbols in order to be able to present the maximum amount of information in as condensed a manner as possible. Weather stations all over the world supply observational data and measurements for the compilation of these weather maps. This compilation was formerly a wearisome task, but today the work is almost totally automated, employing tapes and microfilms to store the data.

The information gathered consists of recorded measurements and observational data on the meteorological elements – air pressure, air temperature, humidity and wind – together with information on the amount of cloud cover and on the forms of precipitation. The individual weather watch and observation stations include information on the following phenomena in their daily reports:

cloud cover (0/8 to 8/8)
type of cloud
height of cloud (cloud base)
air pressure
air pressure change in the last three hours
air pressure tendency (rising or falling)
air temperature
wind direction
wind strength
visibility
dewpoint
amount of rainfall in last six hours
weather over last hour
weather over last six hours

The information in reports from weather stations is plotted as a series of symbols. Each station reports on weather conditions at fixed times. Today most countries have a meteorological service and a central office, where the data are assembled and weather maps prepared, and to which observations and measurements from the individual stations are sent every three hours in a coded form by telephone, telegram, radiotelephone or teleprinter. In Great Britain, the central office is the Meteorological Office, Bracknell, Berkshire; in the United States it is the US Weather Bureau, Washington. A World Meteorological Organization (WMO) has been established to help in the exchange of information between national organizations to fill the gaps not covered by national observation stations. It is normal to use an international code in reporting the

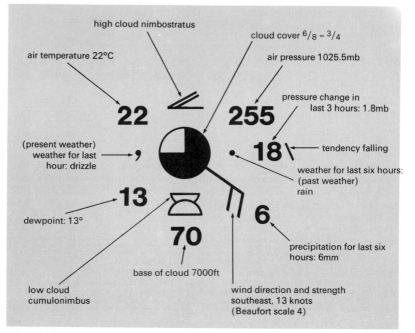

high cloud nimbostratus

cloud cover 6/8 = 3/4

air temperature 22°C

air pressure 1025.5mb

pressure change in
last 3 hours: 1.8mb

22 **255**

(present weather)
weather for last ⟶ **,**
hour: drizzle

18 ⟵ tendency falling

weather for last six hours:
(past weather)
rain

13

dewpoint: 13°

70

6

precipitation for last six
hours: 6mm

base of cloud 7000ft

low cloud
cumulonimbus

wind direction and strength
southeast, 13 knots
(Beaufort scale 4)

A typical meteorological station's report

observations at individual stations, the report being completed with numerical values, in five groups.

The various national meteorological services employ a standard form for reporting observations – for Britain, this is laid down in the *Handbook for Observers*, issued by the Meteorological Office.

The codification of the information facilitates the relatively rapid preparation of weather maps, even at sea. At first sight, a weather map, with all the observations recorded on it, appears quite chaotic – a jumble of symbols, figures and wind arrows. With the help of a key, it is possible to understand the reports of individual stations (see p. 124), but this is still a long way from the science or art of making forecasts. The meteorologist must begin by analysing the data and, as the result of the analysis obtained, prepare an easily comprehensible outline of:

1. The pressure field – as shown by the pattern of isobars
2. The position of fronts and the boundaries of air masses – marking the temperature field.

This systematic ordering of the data enables a physical explanation to be made of the state of the weather and helps to clarify the relationship between individual phenomena.

The data from weather stations are used to prepare a ground-level weather map. The information for upper-air weather maps is provided by radio sondes. The analysis of upper-air conditions is made by the topographical presentation of pressure surfaces (for example, altitudes of the 500mb surface). This gives a view of the pressure and air movements in the free atmosphere. In the analysis of the ground-level weather maps, more attention is paid to fronts and air masses.

The weather map is, however, a fleeting view, the position at a given moment in time, for the sequence of weather continues. With the help of a weather map and a

A simplified weather map for 5 August, 1949, 18.00 hours, designed to assist in teaching. Explanations of the symbols are given in the diagrams, pp. 122 and 124. Shaded area = precipitation (green on coloured maps); solid arrows = cold air (blue on coloured maps); hollow arrows = warm air. (Cold fronts would be shown in blue, warm in red, and occluded in violet).

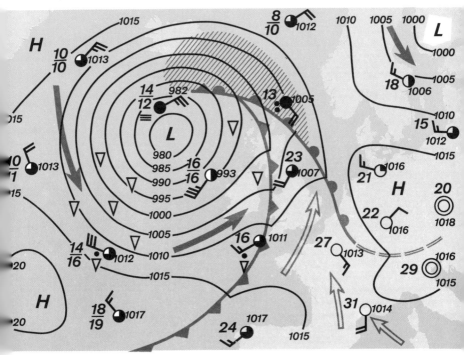

The symbols used on a weather map

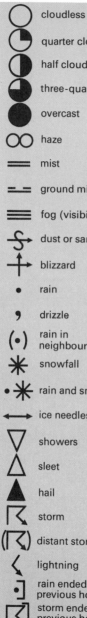

cloudless

quarter cloud cover

half cloud cover

three-quarter cloud cover

overcast

haze

mist

ground mist

fog (visibility under 1 km)

dust or sandstorm

blizzard

rain

drizzle

rain in neighbourhood

snowfall

rain and snow

ice needles

showers

sleet

hail

storm

distant storm

lightning

rain ended in previous hour

storm ended in previous hour

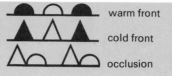

warm front

cold front

occlusion

Windspeed

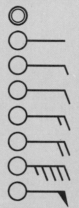

calm or very light

up to 1m/sec. (1–5km/hr)

up to 2.5m/sec (6–13km/hr)

up to 5m/sec (14–22km/hr)

up to 7.5m/sec (23–31km/hr)

up to 10m/sec (32–40km/hr)

up to 22.5m/sec (77–85km/hr)

up to 25m/sec (86–94km/hr)

Cloud symbols

cirrus (ice clouds, feathery)

cirrostratus (ice clouds, as a veil)

cirrocumulus (ice clouds)

altostratus (thick sheets, cloud of ice and water)

nimbostratus (thick sheets, cloud of ice and water)

altocumulus (water)

stratocumulus (water)

stratus (high fog-like closed cloud)

cumulus

cumulonimbus (rain and shower cloud)

altocumulus – (lenticularis lens-like clouds of cooled water droplets – Föhn clouds)

knowledge of the preceding weather pattern, the meteorologist can determine what changes can be anticipated for a given period of time.

On the basis of these changes, a forecast chart is prepared and a weather forecast formulated. For the forecast two methods are possible: (1) forecast based on synoptic laws (2) forecast with the help of mathematical formulae.

Synoptic means a general or collective view. The synoptic laws are based on practical experience, simple theoretical relationships and the field distribution of various meteorological elements (see p. 95), for example, the distribution and movement of pressure systems. To this must be added the positions of fronts and the dominant air movements. The position of upper winds has to be taken into consideration, as with, for example, jet streams. There are many difficulties, for changes in the pressure field lead to changes in wind movement and, thus, to temperature change and air-mass movement. Since observation stations do not always cover interlocking areas, the data available may tend to concentrate on selected areas, with gaps appearing elsewhere. These lead to false evaluations and mistakes in forecasting. It must therefore repeatedly be emphasized that weather prognoses with a synoptic approach, even at the best of times never exceeds a success ratio of 4:1. The question remains as to whether this will change with the substitution of mathematical formulas for experience? Because physical processes are governed by laws, it is possible to consider prognoses using mathematical formulae, but only since the introduction of computers has it been feasible economically in terms of the time available to carry out the forecast. Meteorologists have achieved good results by mathematical means in dealing with prognoses of wind fields in the upper atmosphere. Such prognosis has become essential to weather forecasting, and it should be noted, has led to more accurate results. Important, too, are the radio sondes released twice daily from some stations. In the northern hemisphere they are released from some 100 weather stations, based in a wide range of countries in Europe, Asia, North America and Africa. Radio sondes, as outlined on p. 108, yield data on upper winds, pressure, temperature and humidity. The following information can then be obtained:

height of $0°$
height at which cumulus cloud begins (condensation level)
ground temperature required to allow convective ascent
upper and lower cloud limits
lower and upper inversion levels
windspeed and direction at given heights
boundary between troposphere and stratosphere

Weather reports 23 December, 1976 at 07.00 hrs

station	height a s l	weather	wind in km/hr	temperature °C at 2m.	dew point	max. yesterday	average yesterday	min. at night 2m	grass	sunshine in hours yesterday	rain in 24 hours mm	snow height cm
Berlin	50	overcast	calm	-1	-1	4	1.6	-3	-5	0.5	—	
Schleswig	43	overcast	NNW 7	1	1	3	2.2	1	1	—	—	remnants
Hamburg	16	fog	WSW 5	1	1	5	2.8	0	-3	2.7	—	1
Bremen	3	fog	calm	-2	-2	6	3.0	-3	-4	2.3	—	1
Hannover	55	cloudless	ENE 2	-4	-4	6	2.6	-5	-7	0.7	—	
Essen	154	cloudless	ESE 7	-2	-2	4	1.7	-2	-5	2.8	—	

Other European stations

station	weather	wind in km/hr	temperature °C 07.00 hrs	maximum yesterday
Athens	overcast	N 29	10	12
Bucharest	overcast	NNE 7	1	1
Copenhagen	overcast	ENE 14	2	4
London	overcast	ESE 9	6	10
Madrid	fog	calm	2	2
Moscow	snow	NSW 4	-4	0
Paris	overcast	ESE 4	6	7
Rome	overcast	NE 9	12	15
Stockholm	overcast	NW 9	-4	0
Warsaw	overcast	WSW 4	1	2
Vienna	overcast	SE 16	0	2
Zürich	fog	NE 4	-3	-3

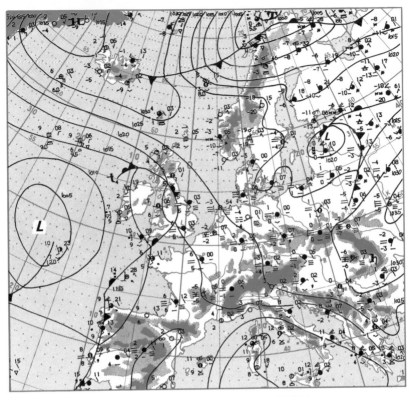

Section of a weather map, 23 December, 1976 at 07.00 hrs

This weather map is the result of all the weather observations made at 07.00 hours at stations belonging to the German meteorological service and reports from foreign stations in international code, made by radio and teleprinter to the central office. There are also electronic plotters that receive information by punched strip and are computer-guided. The symbols used are explained on p. 124. If the weather map is in colour, then the warm front is in red, the cold front in blue, the occlusion in violet. The lines in black are isobars, which connect the points on the earth's surface having equal barometric pressure at the specified time. Dashed black lines (which do not appear on this map) would indicate troughs or convergences. If it were present, rain would be indicated by green shading (or dots); dark green shading or stars would indicate snow; yellow shading or dashes would indicate fog. With the help of this map the weather situation was described as follows: 'Germany remains in a zone of weak pressure between a high over the central Soviet Union and a high centred over the Strait of Denmark. Cold air persists over the ground and determines the weather, but warm air is present in the mountains. Further outlook: the cold front behind the depression over the North Cape in the extreme north of Norway will move southward, and influence the weather in the area from the second day of the Christmas holiday.'

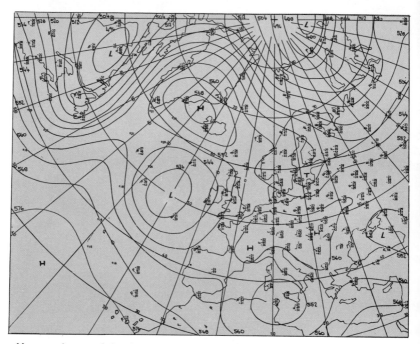

Upper-air map (absolute topography 500mb, or approximately 5500m), 23 December, 1976, 01.00 hrs

In contrast to the weather map on p. 127 showing the conditions at ground level and emphasizing air masses and fronts, this upper-air map provides essential additional information on events in the higher atmosphere, in particular air pressure and air movement. The map is prepared from data supplied by radio sondes.

Topography does not have the same influence on atmospheric weather, so mountains, etc, are not marked and the isohypses (contour lines of a given air pressure) have generally a smoother run than do the isobars at ground level. It should be reiterated that weather events at the ground and at high altitudes are not in step – although weather events at the ground affect those above, and vice versa.

The upper-air chart is in no way a projection of the ground level chart. The form it takes can in fact be quite different. The development of depressions at ground level is followed more slowly at altitude. There may well be a high pressure system in the upper atmosphere, above a depression at ground level with a strongly-developed warm sector. The centre of a high pressure system at altitude is situated northwest of the centre of a depression at ground level. The cirrus clouds at heights are indicative of the air movements.

Significant in the examination of upper-air charts are the so-called upper depressions, often accompanied by an extreme inbreak of cold air. In summer, particularly, cold air aloft in association with warm air at ground level makes for a labile (unstable) weather system, manifested by showers and heavy rain.

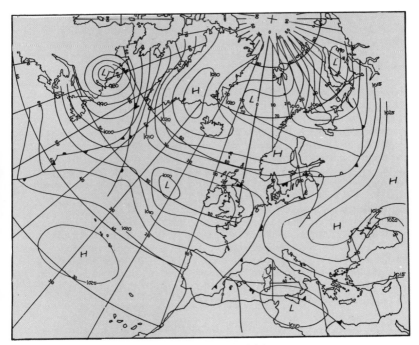

Forecast chart for 23 December, 1976, 01.00 hrs

The station reports given on p. 126 form part of the basis of the weather map provided by the German weather service on p. 127. A valuable addition to this is provided by the measurements of radio sondes and the upper-air map prepared from these measurements (p. 128). A major task of the meteorologist is now to provide forecasts based on these charts and, for this purpose, a forecast chart is constructed. (It is *important to note* that the forecast chart for 23 December, 1976, 01.00 hours, is based on observations and analyses made at 07.00 hours on the previous day, 22 December.)

The forecast chart shows the expected distribution of pressure, the location of fronts and the change in temperature – on these values the forecast is built. This forecast chart can be compared with the actual chart for 23 December, 1976, on page 127, although it can be seen that the forecast chart shows conditions 6 hours earlier than does the actual chart.

The depression has moved from west to east very quickly, covering some 800km (500mi) in 24 hours. This speed decreases as cold and warm fronts intersect, and the cyclone nears the occlusion stage.

The forecast chart would be supplemented by forecasts for individual regions within countries and by reference to the further outlook.

The computer has made mathematical forecasting possible and, similarly, weather satellites have improved measurement and observation data for individual weather phenomena. The comprehension of cloud structures over large areas – such as in the area of fronts, or in the initial stages of a cyclone – are only possible with weather satellites.

Modern forecasts combine the synoptic and the mathematical approaches. It must be kept in mind at all times that the processes involved in weather are very complex. Despite improvements in the technical equipment, the success rate remains limited. A 100 per cent success in forecasting must be regarded as a utopian dream. Indeed, meteorologists would be more than pleased with 85–90 per cent success rate.

Weather maps and forecasts are disseminated daily by various news media (TV, newspapers, radio, telephone, etc.). The maps published by the media are usually of ground level conditions. They contain meteorological symbols indicating pressure systems and fronts (in simplified form), together with wind direction and strengths. A knowledge of these symbols is essential and, to facilitate the reader's task, a key is provided on p.124.

The weather maps for any extensive region in middle and high latitudes show well-defined pressure systems, which extend generally over several thousand kilometres. These systems are represented by isobars (lines or contours connecting geographical points with the same pressure at sea level). The winds blow very nearly along the isobars with the low pressure to the left – in the southern hemisphere, it is to the right. The principal forms that these pressure systems take are:

1. Areas of low pressure (lows) – surrounded by closed isobars; also known as depressions or cyclones.
2. Areas of high pressure (highs) – surrounded by closed isobars; also known as anticyclones.
3. Troughs of low pressure – along which the direction of accompanying winds changes gradually or abruptly; responsible for changing the direction of isobars.
4. Wedges of high pressure – along which the direction of accompanying winds changes gradually; responsible for changing the direction of isobars.

Fundamental to weather observation, however, is the study of the general patterns in which these systems of pressure normally are distributed. In Europe these patterns have been observed for several centuries, making it possible to distinguish a series of typical patterns, knowledge of which makes the forecaster's life considerably easier. Analysis and assessment of these types is based on the occurrence – in terms of time and place – of warm and cold fronts. Fronts are differentiated by the sequence of warmer and colder air – a warm front being one along which warmer air replaces colder air, and a cold front being one along which colder air replaces warmer air.

For our purposes, we shall further typify systems by using the terms westerly, northerly, southerly and easterly. In the next section of the book a range of typical systems and conditions for Europe is described. Opposite each description will be the ground level forecast chart which would cover such conditions, together with an inset showing the upper-air chart for 500 millibars. Whereas on the ground level weather maps the pressure is given in isobars, on the upper-air inset the pressure is given in contour lines (isohypses) of a selected pressure surface. The 500 millibars surface corresponds to an altitude of about 5500 metres (18,000ft). Below that surface is about half the atmosphere of immediate relevance to the study of weather. The movements at 500 millibars provide valuable indications of the movements of highs and depressions at ground level. The upper-air maps also include wind directions and temperatures, as well as the isohypses for pressure distribution.

One further word is necessary on personal observations with a weather map at hand. The more frequent the lift of warm air masses and of cold air inbreaks, the more frequent the weather changes and the more difficult it is to forecast. Westerly conditions, with a rapid sequence of depressions, occur during all seasons, and are characteristic of Britain and much of Europe – they may contain wedges of high pressure or cold fronts bringing interludes of better weather (see pp. 56 and 44). The safest prognoses can be made for weather associated with stationary high pressure systems, extending to the upper layers of the troposphere and covering large parts of the Continent, for example, during autumn and winter over Scandinavia and eastern Europe.

Typical European weather systems and conditions

Anticyclonic westerly system over central Europe
31 January, 1968

Weather type. Westerly

Circulation form. Zonal – a west-east flow between a subtropical high over the north Atlantic, and a depression in the subpolar region. In this airflow, a series of depressions, with associated fronts, moves west-east from the north Atlantic over the European continent.

Conditions. These are characterized by individual disturbances that wander from the north Atlantic over Scotland and southern Scandinavia to northwestern USSR. Outlying fronts are only temporarily present in central Europe and only mildly effective. The central low pressure area is usually in the vicinity of Iceland. The subtropical high (Azores high) lies north of the Azores, but a wedge from it stretches across Spain to southern central Europe.

Weather. Over the British Isles and the North Sea, southern Scandinavia and the north German lowlands there is usually, both in summer and winter, changeable and relatively mild weather. Even in winter the precipitation is mainly as rain, although there is always the possibility of ice. Strong winds often occur, mainly southwesterly. In the central mountains of Germany, such as in the Black Forest region, there is some snowfall in winter. Frontal storms occur during the summer. Again in the central mountain area of Germany, radiation frost is a risk in winter, but in summer the weather can be relatively warm. With these weather conditions, the tendency generally is towards open weather from the British Isles across central Europe to southern central Europe; along the Alps, however, heavy rain and storms take place, usually with moderate southwesterly and westerly winds. Further south, in Spain and Italy, the weather is predominantly fine.

Most frequent occurrence. August and September.

Least frequent occurrence. December and March.

Related weather systems. A cyclonic westerly system over central Europe (see p. 134) and a ridge of high pressure over central Europe, linking the Azores high with a high over eastern Europe (see also high over North Sea and Scandinavia, p. 156).

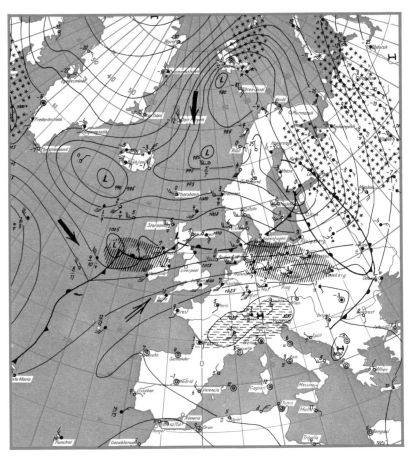

Above: ground level weather map; for explanation of symbols, see p. 124

Right: upper-air map
Lines = contours (isohypses) of the 500mb surface, interval 12 decametres (394ft)
Arrows in the lines = wind direction
Temperature of the 500mb surface = −28°
H = high L = low

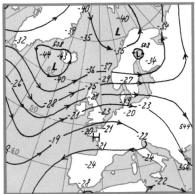

Cyclonic westerly system over central Europe
15 January, 1968

Weather type. Westerly.

Circulation form. Zonal – a west-east flow between a subtropical high over the north Atlantic and a depression in the subpolar region. In this airflow a series of depressions, with associated fronts, move west-east from the north Atlantic over the European continent.

Conditions. Disturbances with alternating high pressure move from the north Atlantic in latitudes 50°–60° over the British Isles, North Sea and the Baltic into eastern Europe. In winter the fronts over eastern Europe move off in a north-east direction. The centre of the depression is located in the north Atlantic near Iceland. Outliers of the Azores high reach Spain and southern France, occasionally even the Alps.

Weather. The weather in north, west and central Europe is completely changeable, alternating between showers and longer periods of rain (in winter the showers possibly beginning with snow), with warmer periods of a few hours or a day under the influence of a high pressure interval. The wind is strong to stormy, and westerly. In winter the associated weather is relatively mild, in summer cool with occasional frontal storms. In central Europe the rainfall diminishes towards the south and east. In the Mediterranean and over the Balkans, the weather is good – dry with little wind.

Most frequent occurrence. July and August.

Least frequent occurrence. March and April.

Related weather conditions. Anticyclonic westerly conditions in mid-latitudes (see p. 132) and a westerly system in southern Europe (see p. 136).

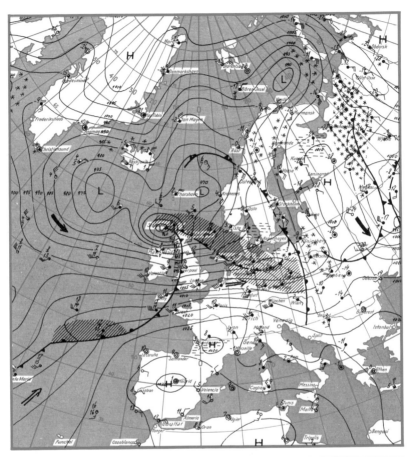

Above: ground level weather map; for explanation of symbols, see p. 124

Right: upper-air map
Lines = contours (isohypses) of the 500mb surface, interval every 12 decametres (394ft)
Arrows in the lines = wind direction
Temperature of the 500mb surface = −28°C
H = high L = low

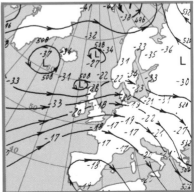

Westerly system in southern Europe
25 May, 1967

Weather type. Westerly.

Circulation form. Zonal – a westerly airstream between a subtropical high over the north Atlantic and a low in the subpolar regions. In this airstream a series of depressions, with associated fronts, move from the north Atlantic into Europe.

Conditions. The frontal zone has moved markedly south. The depressions follow a trajectory south of the British Isles over France, and southern Germany into eastern Europe. From there the depressions move northeastwards. Cyclonic influences penetrate to northern Italy and the northern parts of the Mediterranean region. The central low, controlling the movement, is usually situated northwest of the British Isles, so that a part of the north Atlantic and the North Sea are influenced by the polar high. From this high comes easterly air. The Azores high is effective only as far as northern Africa.

Weather. In central Europe and western Europe, skies are cloudy and the precipitation is plentiful. In spring it is warm or even oppressive, but in summer it is cool by contrast. In northern Germany and on the coasts of the Baltic, the chances of snow are increased by cold air moving from the northeast. Rain occurs in southern France and northern Italy. Sunny, dry weather predominates in Spain and northern Africa. Over western and central Europe, winds are fresh and from the southwest.

Most frequent occurrence. December and March.

Least frequent occurrence. May and September.

Related weather conditions. Cyclonic westerly conditions over central Europe (see p. 134). Highs over Scandinavia (see p. 154), and over the North Sea and Scandinavia (see p. 158).

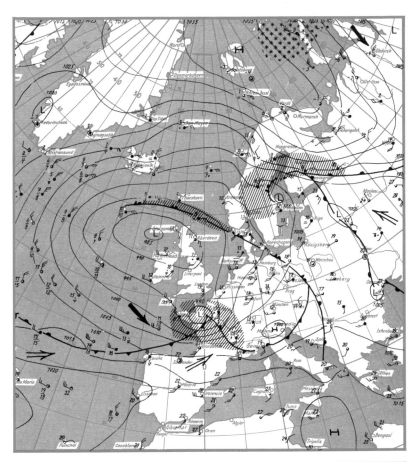

Above: ground level weather map; for explanation of symbols, see p. 124.

Right: upper-air map
Lines = contours (isohypses) of the 500mb surface, interval every 12 deca-metres (394ft)
Arrows in the lines = wind direction
Temperature of the 500mb surface = −28°C
H = high L = low

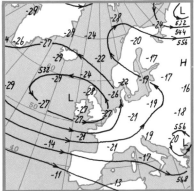

Angular westerly system
27 November, 1967

Weather type. Westerly.

Circulation form. Zonal – a west-east airstream between a subtropical high over the north Atlantic and a depression in the subpolar region. In this airstream a succession of depressions, with associated fronts, move east from the north Atlantic into Europe.

Conditions. A strongly-developed frontal zone in the vincity of Iceland and the British Isles turns sharply northwards as it reaches mainland Europe, caused by the presence of high pressure over the USSR. The bad weather coming from the Atlantic reaches central Europe only, the disturbances there becoming stationary. The Soviet high pressure dominates in eastern Europe and the Balkans.

Weather. In northern and western Europe, and in parts of central Europe, weather is changeable with rain, sometimes very heavy in France and West Germany. In winter there is snowfall in Scandinavia and central Europe. There is no precipitation, however, in eastern Europe because of the influence of the Russian high, and in winter the weather is cold. Southern France is cloudy with perhaps some rain, with similar conditions in north Italy and parts of the western Mediterranean region. In summer, temperatures in western and central Europe associated with this system are only moderate. A stiff breeze blows from the west.

Most frequent occurrence. November and December.

Least frequent occurrence. May and July.

Related weather conditions. In central Europe predominantly cyclonic southeasterly conditions giving way to a high over central and southern USSR. In central Europe, particularly in March and October, anticyclonic southwesterly with dry weather and early fog in spring and autumn. A related weather system is the high over Scandinavia (see p. 152).

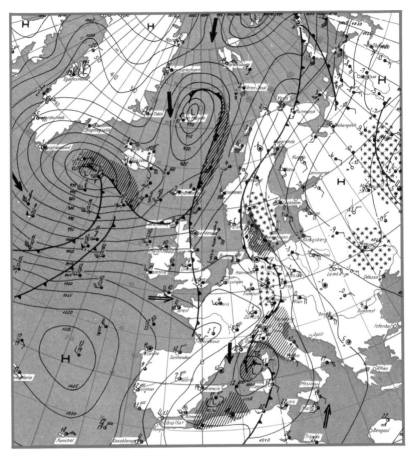

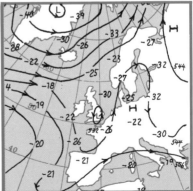

Above: ground level weather map; for explanation of symbols, see p. 124.

Right: upper-air map
Lines = contours (isohypses) of the 500mb surface, interval every 12 decametres (394ft)
Arrows in the lines = wind direction
Temperature of the 500mb surface = −28°C
H = high   L = low

Anticyclonic northerly over central Europe
12 January, 1968

Weather type. Northerly.

Circulation form. Meridional – a stationary high ('blocking' high) is particularly effective in promoting this form of air movement. Northerly conditions in the high are over the British Isles while a depression is over the Baltic Sea and northwestern USSR.

Conditions. A high pressure system stretches north-south over the British Isles and the North Sea. Occasionally, this system establishes itself from Spain to Spitzbergen. A well-developed depression is centred over eastern Europe. The wedge of high pressure over the British Isles reaches high latitudes. On the eastern flanks of the high depressions wander southwards and southeastwards.

Weather. Norway and western Europe are under the influence of the high pressure system, bright or cloudy, with occasional snow showers in winter. During clear weather, frosts occur. Rainfall increases eastwards, with heavy snowfalls in winter and, occasionally, deep frost. In summer the weather is relatively cool in central Europe, in east central Europe and eastern Europe it is markedly cold. The depression over eastern Europe is associated with stormy winds from the north and northwest. Freshening winds from this direction are present in central Europe, France and Italy.

Most frequent occurrence. May and June.

Least frequent occurrence. January and October.

Related weather conditions. A cyclonic northerly over central Europe (see p. 142), a high over the British Isles (see p. 148) and an anticyclone northwest of central Europe with a high on the western margins of the continent and a depression over Scandinavia and the North Sea.

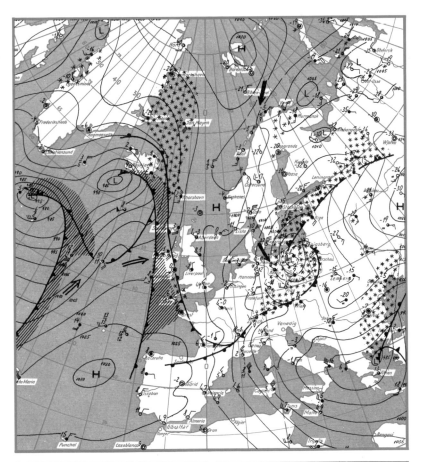

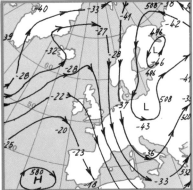

Above: ground level weather map; for explanation of symbols, see p. 124.

Right: upper-air map
Lines = contours (isohypses) of the 500mb surface, interval every 12 deca-metres (394ft)
Arrows in the lines = wind direction
Temperature of the 500mb surface = –28°C
H = high L = low

Anticyclone west of British Isles
(Cyclonic northerly over central Europe)
11 March, 1968

Weather type. Northerly.

Circulation form. Meridional – the presence of a 'blocking' or stationary high being particularly effective in promoting this type of air movement. This high is situated between 50° and 65° north, generally over the British Isles, while a depression dominates the Baltic Sea and northwest Russia.

Conditions. An extensive high is west of the British Isles, occasionally with a ridge of high pressure linking a high over Spain with a high over Spitzbergen. The depression is present over Scandinavia and western USSR. Between these two pressure systems, cold maritime air moves southwards to central and western Europe. Bad weather extends from Iceland over the North Sea to the south and southeast. The cold air penetrates the Mediterranean region, there generating disturbances.

Weather. All Scandinavia, eastern and central Europe, much of western Europe, Spain and Italy have variable weather with rain. Winds are strong, sometimes squally and from the north. The rainfall in the mountains is very heavy. There are snowfalls in winter and spring, in central Europe particularly there being many rain and snow showers (April weather). In spring it is often 'cold for the time of year'. This system brings colder than normal weather during all seasons of the year.

Most frequent occurrence. May and June.

Least frequent occurrence. November and December.

Related weather conditions. A trough over central Europe (see p. 150), a cyclonic system northwest of central Europe with strong frontal zone extending over the British Isles, North Sea and eastern central Europe, an anticyclone north of central Europe (see p. 140) and a high over the British Isles (see p. 148).

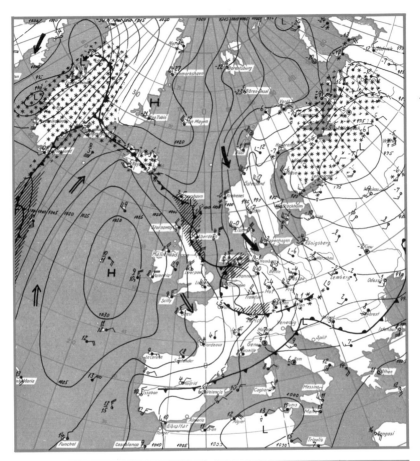

Above: ground level weather map; for explanation of symbols, see p. 124.

Right: upper-air map
Lines = contours (isohypses) of the 500mb surface, interval every 12 decametres (394ft)
Arrows in the contours = direction of wind
Temperature of the 500mb surface = −28°C
H = high L = low

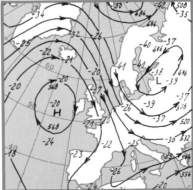

High over the North Sea and Iceland
(Anticyclone over central Europe)
27 May, 1966

Weather type. Northerly.

Circulation form. Meridional – a stationary or blocking high is particularly effective in promoting this form of air movement. The high is situated between 50° and 65° north, over the British Isles. A depression lies over the Baltic Sea and northwestern USSR.

Conditions. In this case the blocking high is situated far north over the North Sea as far south as Scotland. A wedge of high pressure extends to central Europe. A west Atlantic frontal zone is sometimes turned north (over Greenland and further northeast) and sometimes south (southwest Europe to the Mediterranean). The occasional extensions of the depression in the USSR southwards only affect eastern Europe.

Weather. Over west and central Europe, weather is mostly warm and cloudy. Rainfall (associated with storms) occurs in east central Europe and even more so in eastern Europe. In winter this pattern brings hard frosts (radiation frost) to central Europe, and ground frost can also come in spring. Clear skies in summer brings fine warm weather, but without intense heat. In the Mediterranean region, particularly in the eastern and central parts, rainfall can occur, resulting from low pressure and incoming depressions. Winds over the Mediterranean are moderate, whereas over the North Sea they are fresh to strong, and northerly or northeasterly.

Most frequent occurrence. May and June.

Least frequent occurrence. February and November.

Related weather conditions. High over the North Sea and Iceland, cyclone over central Europe (see p. 146), cyclonic northerly situated over central Europe (see p. 142), anticyclonic northerly over central Europe (see p. 140) and a high over the British Isles (see p. 148).

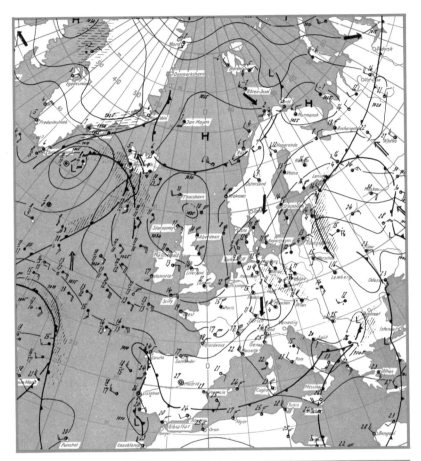

Above: ground level weather map; for explanation of symbols, see p. 124.

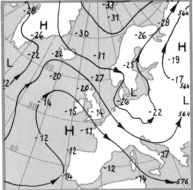

Right: upper-air map
Lines = contours (isohypses) of the 500mb surface, interval every 12 deca-metres (394ft)
Arrows in the contours = direction of wind
Temperature of the 500mb surface = −28°C
H = high L = low

High over the North Sea and Iceland
(Cyclone over central Europe)
22 August, 1966

Weather type. Northerly.

Circulation form. Meridional – a stationary or blocking high is particularly effective in promoting this form of air movement. The high is situated between 50° and 65° north, over the British Isles. A depression is centred over the Baltic Sea and northwestern USSR.

Conditions. The air movements and pressure distribution resemble those described on p. 144 with a high over the North Sea and Iceland. The wedge of high pressure extending over the North Sea to west central Europe is missing. There is therefore greater opportunity for the frontal zone to move into southwestern Europe and the Mediterranean region, influencing the weather of central Europe more effectively. Also, cold air from the Baltic Sea can generate disturbances in central Europe.

Weather. Rainy weather predominates in south central Europe and western Europe – in winter as snow, in summer as storms with very heavy showers. Whereas in western Europe and in south central Europe, the weather associated with this system can be relatively mild, in the northern parts of Central Europe, in East Europe and Scandinavia the conditions in winter are very cold and in summer cool. Better clearer weather is present in Britain and Iceland, in Spain and in Portugal. Southern France and Italy are changeably cloudy with some rain.

Most frequent occurrence. May and October.

Least frequent occurrence. August and September.

Related weather conditions. Cyclonic northerly over central Europe (see p. 142), westerly conditions in southern Europe (see p. 136), isolated depression over central Europe, a high over North Sea and Iceland, with an anticyclone over central Europe (see p. 144).

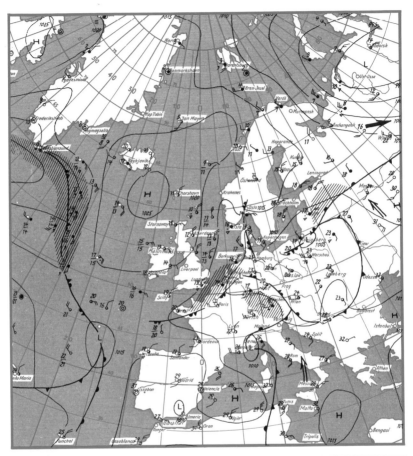

Above: ground level weather map; for explanation of symbols, see p. 124.

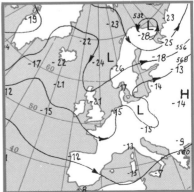

Right: upper-air map
Lines = contours (isohypses) of the 500mb surface, interval every 12 decametres (394ft)
Arrows in contours = direction of wind
Temperature of the 500mb surface = −28°C
H = high L = low

High over the British Isles
8 April, 1968

Weather type. Northerly.

Circulation form. Meridional – a stationary or blocking high is particularly effective in promoting this form of air movement. The high is situated between 50° and 65° north, over the British Isles. A depression is centred over the Baltic Sea and northwestern USSR.

Conditions. A blocking high is situated directly over the British Isles or, immediately adjacent, over either the North Sea or Atlantic Ocean. Often this high is connected with a high pressure area over Greenland and Iceland. The depression is generally over the Baltic Sea, usually reaching far to the southeast. Marginal disturbances between the high and the depression influence the weather of all central Europe. Often the pressure is low over the Mediterranean.

Weather. The airstream over central and western Europe is mainly northerly. In summer the weather is mostly dry and sunny, but in winter a thick layer of fog can form above ground level with dull weather. Clouds thicken towards east central Europe, and rain occasionally occurs, possibly heavy in the eastern Alps. At all seasons only low temperatures are experienced, with moderate to deep frost in the winter months. South of the Alps the weather is changeable, particularly in Italy.

Most frequent occurrence. February and September.

Least frequent occurrence. July and December.

Related weather conditions. High pressure over the North Sea and Iceland, with an anticyclone over central Europe (see p. 144). High pressure over the North Sea and Iceland, with a cyclone over central Europe (see p. 146); cyclonic and anticyclonic northwesterlies, with changeable weather and some heavy rain on the northern side of the Alps.

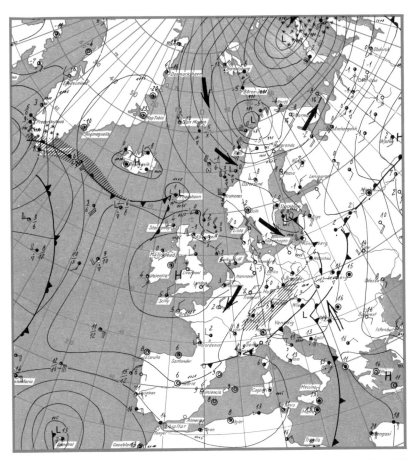

Above: ground level weather map; for explanation of symbols, see p. 124.

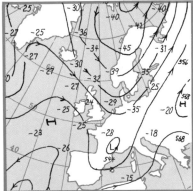

Right: upper-air map
Lines = contours (isohypses) of the 500mb surface, interval every 12 deca-metres (394ft)
Arrows in contours = direction of wind
Temperature of the 500mb surface = −28°C
H = high L = low

Trough over central Europe
31 December, 1967

Weather type. Northerly.

Circulation form. Meridional – a stationary or blocking high is particularly effective in promoting this form of air movement. The high is situated between 50° and 65° north, over the British Isles. A depression is centred over the Baltic Sea and northwestern USSR.

Conditions. Particularly low pressure far behind the cold front of a depression, indicates a trough influencing northern and central Europe. The eastern north Atlantic and the northern part of the Baltic region are under a high pressure. Disturbances move along the frontal zone over western and central Europe, remaining active into the Mediterranean area, particularly in its eastern parts.

Weather. With fresh westerly and northerly winds, the weather of western and central Europe is characterized by great changeability. Showers (of snow in winter) are frequent. In western and central Europe this situation is associated with low temperatures during all seasons, in contrast to eastern Europe where even in winter temperatures are about 0°C or above. The rainfall in eastern Europe is therefore particularly heavy. The Mediterranean region, particularly in its eastern parts, remains under the influence of disturbances coming from western and central Europe.

Most frequent occurrence. April and November.

Least frequent occurrence. June and August.

Related weather conditions. Cyclonic northerly over central Europe (see p. 142) and northwesterly conditions; a cyclone over central Europe, with cold weather and showers (heavy snowfalls in winter).

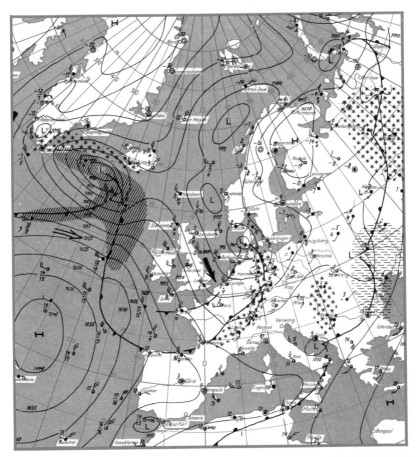

Above: ground level weather map; for explanation of symbols, see p. 124.

Right: upper-air map
Lines = contours (isohypses) of the 500mb surface, interval every 12 decametres (394ft)
Arrows in contours = direction of wind
Temperature of the 500mb surface = −28°C
H = high L = low

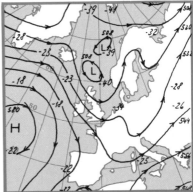

High over Scandinavia
(Anticyclone over central Europe)
9 January, 1966

Weather type. Easterly.

Circulation form. Meridional – a stationary or blocking high is particularly effective in promoting this form of air movement. The high is situated between 50° and 65° north. In its more eastward situation, it centres over northern Europe, while a low pressure system dominates the Mediterranean region.

Conditions. These conditions, particularly frequent in winter, are distinguished by a strongly-developed high pressure system over Finland, often also over northern USSR. The Mediterranean depression is poorly-developed. A large depression is situated over the Atlantic Ocean, west of the British Isles. The high pressure influence reaches from northern Europe to central Europe.

Weather. Normally large parts of central Europe enjoy dry sunny weather, accompanied, however, by much cold air coming from the north. In winter this system is therefore associated with hard frosts, and in spring the weather is cold for the time of year. In summer, by contrast, it is quite hot, but without the threat of storms. Western Europe is under the influence of depressions initiated by the Atlantic low pressure system, with a sequence of rain belts followed by showery conditions. The Mediterranean region is changeable, with variable amounts of cloud.

Most frequent occurrence. January and February.

Least frequent occurrence. June and July.

Related weather conditions. High over Scandinavia but a cyclone over central Europe (see p. 154). High over North Sea and Scandinavia but an anticyclone over central Europe (see p. 156). High over North Sea and Scandinavia, a cyclone over central Europe (see p. 158); northeasterly conditions, a cyclone over central Europe (cool and wet); southeasterly conditions, an anticyclone over central Europe (clear and dry, early fogs in autumn and winter), southeasterly conditions, a cyclone over central Europe (changeable weather).

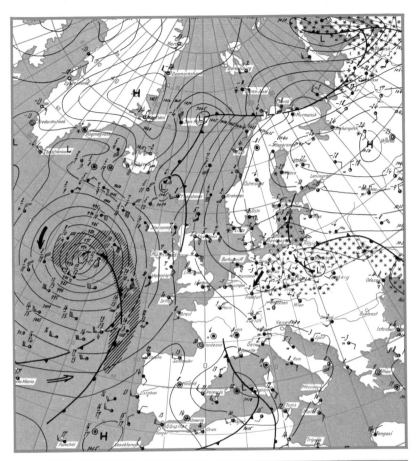

Above: ground level weather map; for explanation of symbols, see p. 124.

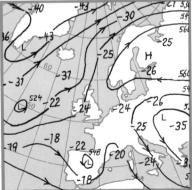

Right: upper-air map
Lines = contours (isohypses) of the 500mb surface, interval every 12 decametres (394ft)
Arrows in contours = direction of wind
Temperature of the 500mb surface = −28°C
H = high L = low

High over Scandinavia
(Cyclone over central Europe)
16 September, 1967

Weather type. Easterly.

Circulation form. Meridional – a stationary or blocking high is particularly effective in promoting this form of air movement. The high is situated between 50° and 65° north, and in its easternmost position is centred over northern Europe, while a depression dominates the Mediterranean region.

Conditions. A high pressure system is centred over Finland, and a low pressure system over the Mediterranean and southern Europe. Between these two pressure systems air moves westwards over central Europe, occasionally reaching the British Isles. Disturbances moving in from the Atlantic are divided, some going north over Greenland to the White Sea, the others over the Bay of Biscay to the Mediterranean. From here, some disturbances reach the southern Alps.

Weather. Clear dry weather in northern Europe, but in contrast variable weather with rain in western Europe, central Europe and in part the Mediterranean region. The precipitation in winter falls mainly as snow in central Europe. In summer these conditions are associated with storms. In winter the weather is cold and frosty, in summer rather oppressive. Winds in central Europe, western Europe and the Mediterranean are easterly.

Most frequent occurrence. March, November, December.

Least frequent occurrence. July, August.

Related weather conditions. High over North Sea and Scandinavia, with a cyclone over central Europe (see p. 158). High over North Sea and Scandinavia, with an anticyclone over central Europe (see p. 156). High over Scandinavia, with an anticyclone over central Europe (see p. 152) and a depression over central Europe.

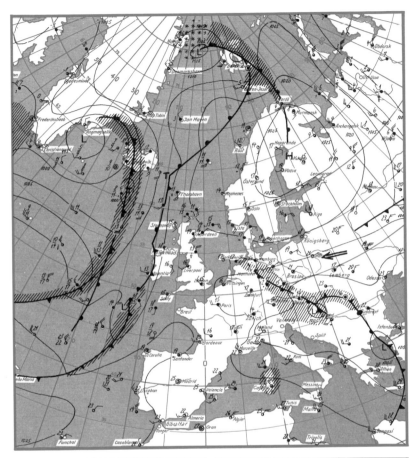

Above: ground level weather map; for explanation of symbols, see p. 124.

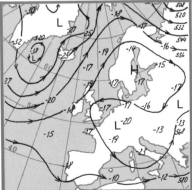

Right: upper-air map
Lines = contours (isohypses) of the 500mb surface, interval every 12 decametres (394ft)
Arrows in contours = direction of wind
Temperature of the 500mb surface = −28°C
H = high L = low

High over North Sea and Scandinavia
(Anticyclone over central Europe)
16 April, 1968

Weather type. Easterly.

Circulation form. Meridional – a stationary or blocking high is particularly effective in promoting this form of air movement. The high is situated between 50° and 65° north, and in its easternmost position is centred over northern Europe, while a depression dominates the Mediterranean region.

Conditions. In this case, the high pressure system extends from Iceland to northern USSR. The northern parts of central Europe are also under its influence. Lower pressures are present over the western Mediterranean region and the southern parts of the North Atlantic. All central Europe and western Europe are under an easterly airstream.

Weather. Northern Europe, a large part of north and east central Europe and western Europe have predominantly clear dry weather. Increasing cloud in south central Europe and south of the Alps is not associated with any greater rainfall. In summer the weather is warm, whereas in winter there is hard frost in north central Europe, North and East Europe.

Most frequent occurrence. May and June.

Least frequent occurrence. July and August.

Related weather conditions. High over North Sea and Scandinavia with a cyclone over central Europe (see p. 158). High over North Sea and Iceland, with an anticyclone over central Europe (see p. 144). High over North Sea and Iceland, with a cyclone over central Europe (see p. 146). Also southeasterly conditions, with a cyclone or an anticyclone over central Europe with predominantly clear dry weather, Föhn winds in the northern Alps (anticyclonic), or changeable with rain, in summer oppressive and stormy (cyclonic).

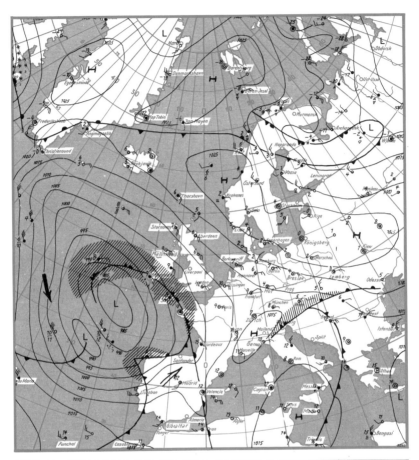

Above: ground level weather map; for explanation of symbols, see p. 124.

Right: upper-air map
Lines = contours (isohypses) of the 500mb surface, interval every 12 deca-metres (394ft)
Arrows in contours = direction of wind
Temperature of the 500mb surface = −28°C
H = high L = low

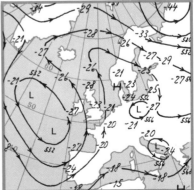

High over North Sea and Scandinavia
(Cyclone over central Europe)
23 February, 1968

Weather type. Easterly.

Circulation form. Meridional – a stationary or blocking high is particularly effective in promoting this form of air movement. The high is situated between 50° and 65° north and in its easternmost position is centred over northern Europe, while the Mediterranean is under the influence of a depression.

Conditions. As when a high is established over the North Sea and Scandinavia, with an anticyclone over central Europe (see p. 156), in this system the high extending from Iceland to northern USSR has a blocking effect. The high is partially divided by the Norwegian coast, and there is very marked cyclonic activity over the western Mediterranean and western Europe. The frontal zone stretching from the west over the Atlantic Ocean is deflected by the high either northeastwards or southeastwards.

Weather. Most of western and central Europe is covered by cloud, with rain (in winter, heavy snow). In summer, this system leads to storms with rain, and much of Europe is warm and oppressive. In northern and eastern Europe, the system is associated with cold wintery weather with hard frost, also in parts of north central Europe. The displaced frontal zone gives the western Mediterranean cloud and rain.

Most frequent occurrence. March and May.

Least frequent occurrence. September, October, December.

Related weather conditions. High over Scandinavia, with a cyclone over central Europe (see p. 154); depression over central Europe, southwesterly system (see p. 136); high over North Sea and Scandinavia, with an anticyclone over central Europe (see p. 156); southeasterly system, an anticyclone over central Europe.

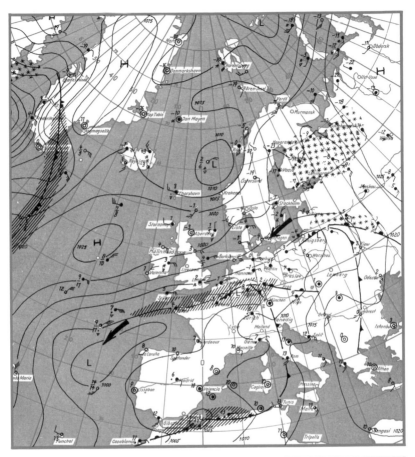

Above: ground level weather map; for explanation of symbols, see p. 124.

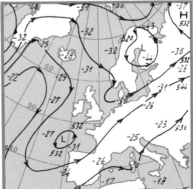

Right: upper-air map
Lines = contours (isohypses) of the 500mb surface, interval every 12 decametres (394ft)
Arrows in contours = direction of wind
Temperature of the 500mb surface = −28°C
H = high L = low

Southerly, high over eastern Europe
(Anticyclone over central Europe)
8 November, 1966

Weather type. Southerly.

Circulation form. Meridional – a stationary or blocking high is particularly effective in promoting this form of air movement. The high is situated between 50° and 65° north. In its southern position it is centred over eastern Europe while a depression dominates western Europe.

Conditions. The high pressure system over eastern Europe is well-developed. A depression is situated over the eastern Atlantic and western Europe. The Atlantic frontal zone runs north of the Azores to the western European coast and then turns northwards. Disturbances are particularly notable in France and Spain.

Weather. In western and southwestern Europe there are occasionally extensive rainbelts. Only the western parts of central Europe are cloudy. In the rest of central Europe clear dry weather prevails, and this is also true of eastern Europe. In this system, mornings in central Europe can be misty, and in winter fog can develop at low heights. In the Alps Föhn winds often develop. In summer weather in central Europe is hot but not generally oppressive. In spring and autumn the weather is mild. Hard frost occurs in winter in eastern Europe and the eastern parts of central Europe (radiation frost). Frost is rare in western Europe – it may be misty.

Most frequent occurrence. January and November.

Least frequent occurrence. June and July.

Related weather conditions. Southerly system, cyclone over central Europe (see p. 162). High over central Europe, southeasterly system.

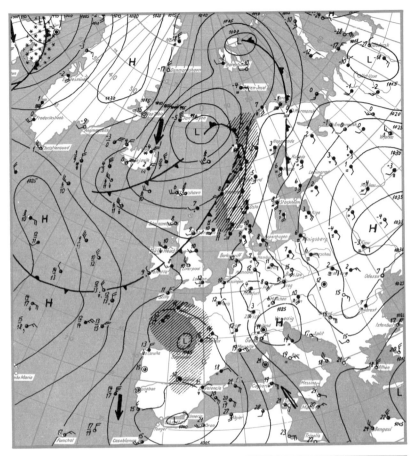

Above: ground level weather map; for explanation of symbols, see p. 124.

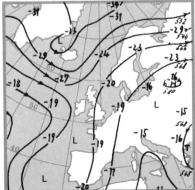

Right: upper-air map
Lines = contours (isohypses) of the 500mb surface, interval every 12 deca-metres (394ft)
Arrows in contours = direction of wind
Temperature of the 500mb surface = −28°C
H = high L = low

Southerly, high over eastern Europe
Cyclone over central Europe
26 February, 1967

Weather type. Southerly.

Circulation form. Meridional – a stationary or blocking high is particularly effective in promoting this form of air movement. The high pressure system between 50° and 65° is centred over eastern Europe in its southerly position. A low pressure system dominates western Europe.

Conditions. The very stable high pressure system of eastern Europe is opposed by the low pressure system between Iceland and the British Isles. The Atlantic frontal zone runs across France and then turns northwards. Disturbances move over France, Spain and Italy in a northerly direction.

Weather. Much of western Europe has cloud and changeable weather. Rainfall amounts are variable. In summer there is a tendency to storms. The precipitation in western and central Europe, even in winter, comes as rain, with a tendency to ground and hill fog. The Alps are under the influence of Föhn winds. Heavy snowfalls occur in Scandinavia and eastern Europe. In western and central Europe temperatures at all seasons are warm, in summer oppressive. Winds in central Europe are southerly, fresh or strong along coasts. Disturbances may also occur in the Mediterranean, with rain in northern Italy.

Most frequent occurrence. February and December.

Least frequent occurrence. May to July.

Related weather conditions. Depression over the British Isles (see p. 164). High over eastern Europe, with an anticyclone over central Europe.

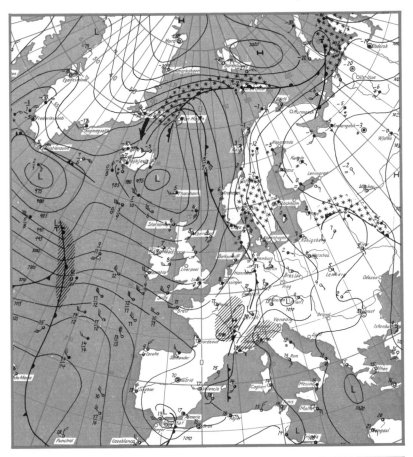

Above: ground level weather map; for explanation of symbols, see p. 124.

Right: upper-air map
Lines = contours (isohypses) of the 500mb surface, interval every 12 deca-metres (394ft)
Arrows in contours = direction of wind
Temperature of the 500mb surface = −28°C
H = high L = low

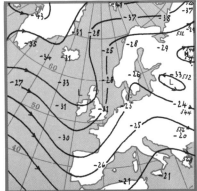

Depression over the British Isles
7 February, 1968

Weather type. Southerly.

Circulation form. Meridional – a stationary or blocking high is particularly effective in promoting this form of air movement. This high pressure system is situated between 50° and 65° north. With this southerly movement over central Europe, the high is centred over eastern Europe; western Europe is dominated by a depression.

Conditions. In this case, the high is very far east in the USSR, with the British Isles dominated by a deep depression. These form a path for depressions moving across France, Italy, central Europe and part of Scandinavia.

Weather. Cloudy with rain belts in the British Isles, France, west central Europe and Italy. The weather throughout is variable and relatively warm, even in winter – indeed, this system brings about thawing conditions in winter. In warmer seasons, storms are possible. Further east the weather is clearer. Föhn winds blow in the Alps. Winds are generally southeasterly in central Europe, and southerly to southwesterly further west and over the British Isles.

Most frequent occurrence. May and August.

Least frequent occurrence. January and March.

Related weather conditions. Southerly, with a cyclone over central Europe (see p. 162); southerly, with an anticyclone over central Europe (see p. 160).

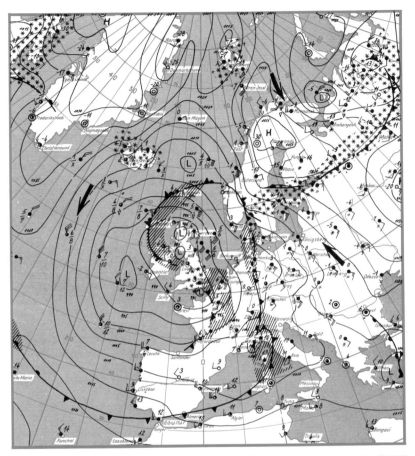

Above: ground level weather map; for explanation of symbols, see p. 124.

Right: upper-air map
Lines = contours (isohypses) of the
500mb surface, interval every 12 deca-
metres (394ft)
Arrows in contours = direction of wind
Temperature of the 500mb surface =
−28°C
H = high L = low

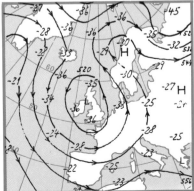

Trough over western Europe
15 November, 1967

Weather type. Southerly.

Circulation form. Meridional – a stationary or blocking high is particularly effective in promoting this form of air movement. With the southerly movement over central Europe, the high is centred over eastern Europe and a depression has been established over western Europe.

Conditions. High pressure over western USSR is opposed by a low pressure system in the form of a trough extending from Spain to the North Sea. The frontal zone, running from the Atlantic towards northern Spain and then northeastwards to central Europe and northern Europe, brings disturbances to those regions.

Weather. Large areas of Europe are cloudy with belts of rain, particularly in northern Spain, France, Germany and the Baltic lands. Storms occur in summer, and the mild temperatures bring the precipitation as rain, even in winter. Föhn winds blow in the Alps. Clearer weather is present further to the east. This is also true of Italy and the eastern Mediterranean, but the western Mediterranean is cloudy – although mainly without rain. Over the Mediterranean, winds are from the south. In western and northern Europe the fresh winds are westerly or northwesterly.

Most frequent occurrence. July and August.

Least frequent occurrence. January and February.

Related weather conditions. Southeasterly conditions.

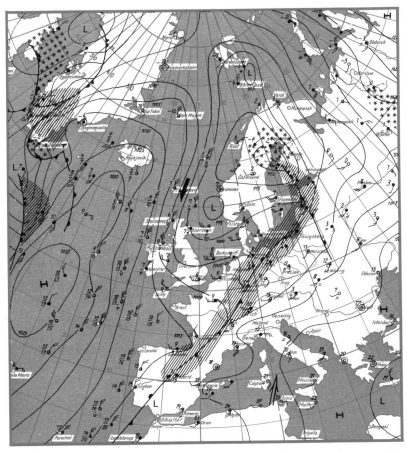

Above: ground level weather map; for explanation of symbols, see p. 124.

Right: upper-air map
Lines = contours (isohypses) of the 500mb surface, interval every 12 decametres (394ft)
Arrows in contours = direction of wind
Temperature of the 500mb surface = −28°C
H = high L = low

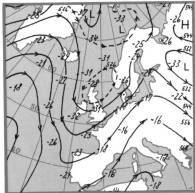

More advanced meteorology

The basic forces at work in the making of weather have been described in the section *What is weather?* p. 13. Having looked at the main weather elements and the range of observable phenomena, as well as having studied with the aid of weather maps a sample of typical weather conditions in Europe, it is now time to return to a more detailed examination of the components of weather.

Heating of land, water and air

Less than half of the radiation from the sun received by the upper atmosphere reaches the earth's surface – most of that received is released into the atmosphere. More radiation is received when the angle of the sun's rays is vertical to the ground (equatorial regions, tropics) than when the sun's rays are at a low angle to the ground (polar regions).

Radiation released by the ground back into the atmosphere is long-wave, and most of it is absorbed when the atmosphere has a high water vapour content. A small amount, about 8 per cent, succeeds in passing back through the atmosphere into space. Clouds hinder the outward radiation from the earth's surface. This constitutes, as we have seen, the 'greenhouse' effect in the atmosphere, comparable with the processes at work in a garden greenhouse. Instead of the glass, it is the water vapour and carbon dioxide in the atmosphere, as well as the water droplets in clouds, that limit the outward radiation.

'Warming from below' is an important aspect of weather, for in its action it promotes vertical air movements. The warming is not achieved solely by conduction from the earth's surface to the atmosphere directly above, for air is a poor conductor of heat. It is much more a question of vertical movements by the air warmed at the surface, these movements transporting the heat to greater heights. In return, cool air from higher in the atmosphere sinks to the earth's surface. This turbulence is basic to the exchange of heat, and mixes air masses of various temperatures.

The different warming rates of land and sea must also be noted. Water reflects more radiation and therefore warms more slowly. The radiation penetrates to about 20 metres (65ft) deep in the water and thus has to warm a greater mass than on the land where the radiation penetrates only a little way into the ground. The land surface also reflects less than the water so that the surface warms more quickly, but it also loses heat more quickly. However, heat is lost above water by evaporation. The different rates of warming of land and sea are discernible in both daily and annual temperature ranges (see also graphs, p. 102). The fundamental processes involved in warming and cooling of land and sea, in the day and night and in a year, are also noted in many weather sayings and proverbs. These processes include the origins of land and sea breezes on the sea coasts and by large lakes, and also of valley and mountain winds (see p. 32). The

The weather in Europe is strongly influenced by the Azores high and the Icelandic low. Warm air from the Azores moves north and, in the region influenced by the Icelandic low, is lifted and cooled. The warm and cold air masses from these two pressure systems give rise to disturbances that move from the Atlantic into Europe.

daily sea breeze occurs not only on the sea coast but, for example, by large lakes such as the Great Lakes of North America or the smaller Lake Geneva in central Europe. During an entire year the different warming and cooling rates of land and sea lead to periodic air movements, known as the 'monsoon effect'. There are in Europe, during June, summer monsoon conditions, with squally northwest winds and wet weather, and, from October, winter monsoon conditions, with dry winds from the southeast. With all weather proverbs, it must be remembered that local circumstances are of great importance. The sequence of weather is quite different in the lowlands from that in the mountains, under the influence of marine climate and under the influence of continental climate (see p. 178).

Over the centuries, many local weather sayings have 'migrated', and are used in areas to which they have never applied. This explains many paradoxes, not to mention faulty forecasts. It also applies to days considered to have some special meteorological significance, such as St Swithin's Day.

As it warms daily in valleys and on slopes, air lifts and cools at 1°C for every 100 metres (328ft). The relative humidity rises and, at a specific altitude, saturation is reached. The resulting condensation produces clouds, at the same time releasing heat, so that at even greater heights cooling is only 0.5°C for every 100 metres (328ft).

Rising and stratified air

Rising air cools – but the rate of cooling depends on the humidity of the air. Dry air on rising cools at 1°C for every 100 metres (328ft). Moist air cools at the same rate, but only until it reaches saturation. If it continues to rise, water vapour condenses and the heat released reduces the rate of cooling to about 0.5°C per 100 metres (328ft). It is quite different with sinking air. Both dry air and moist air gain temperature at 1°C per 100 metres (328ft). This is a process, operating without any external supply of warmth ('adiabatic process'), which is of extreme importance in the daily pattern of weather. The fact, as already mentioned, that pressure falls with height would in itself produce a decrease in temperature with increasing height. Pressure falls with increasing height fairly quickly so that, if air rises quickly, temperatures fall quickly. During certain weather conditions, for example, when a Föhn wind is blowing, a lift of 10 metres (33ft) per second is often achieved. Another example is the so-called 'thermal', in which the air can rise several hundred metres in a few minutes. The air movements in a storm are particularly violent and these reach great heights – as much as 10,000 metres (33,000ft). Cooling at altitude sooner or later leads to cloud formation and, as a corollary, warming with sinking leads to drier air and the absorption of clouds.

The rising and falling of air is not regular and undisturbed. The warming of air takes place first from below. The solar radiation absorbed by the earth's surface is released into the atmosphere with which it comes into contact, but a small part of that radiation is absorbed as it passes through the atmosphere. The atmosphere is thus warmed, a process which is particularly the case in the upper layers. The general rule that temperature decreases with altitude does not always apply. There are stratifications of the atmosphere in which the temperature rises with increasing altitude, in other words, warm air is present above cold air. These particular layers prevent upward and downward movements. The boundary layer between the upper warm air and the lower cold air is known as the inversion level (p. 62).

This boundary layer, acting as a barrier, can be recognized quickly when rising cloud (cumulus cloud) spreads horizontally. It comes as a surprise to many that this type of cloud occurs on days during which the pressure is high, for example, in autumn and winter. Despite the high pressure, the whole day remains cloudy and overcast. The association, generally, of good weather with high pressure is the result of the fact that air moves out from a high pressure system and is replaced from above, the downward movement absorbing the clouds. However, this is not the case when inversion occurs. The vertical movements are terminated and the warming during the day is not sufficient to generate upward movement and break this inversion. The cold layer at ground level is usually only about 100 metres (330ft) thick, the weather being raw and misty. A change takes place during these seasons when the high pressure is replaced by a more labile air mass and the inversion broken up. Above the inversion, the weather is cloudless and visibility excellent.

A stratification with heavy cold air below and light warm air above is relatively stable. Extraordinary labile (unstable) conditions occur when the stratification is composed of heavy cold air above light warm air. The tendency towards turbulent weather is then very great as the cold air attempts to sink beneath the warm, and the warm to rise above the cold. Storms are generally characteristic of such conditions.

Highs and lows

The differences in the heating of land and sea, as well as the relationships between rising and sinking air, makes it possible to understand air movements – and of the frequently employed terms 'high' and 'low'.

With solar radiation throughout the day, land warms more rapidly than an adjacent sheet of water. That means the air over the land is warmed and expands, particularly at altitude. The rising air moves out in all directions, and the pressure over the land falls. A low is formed, and air moves in from all sides. In an isolated model (see also diagram on p. 174), the circulation of air is as follows:

1. Rising air above the heated land surface;
2. at altitude, air moves away and then sinks;
3. at ground level, an inward movement of cool air from above water surfaces;
4. the air sinking from heights replaces the air over the water surfaces and then moves in again to the warmed land surface.

The cycle finally comes to an end when solar radiation stops. At night the land radiates more rapidly than the water. The air over the land cools and contracts. At altitudes, the air is drawn in and increases the total air pressure at ground level. A high pressure is formed. Again a cycle of air movement is formed:

In the low pressure area, air rises after moving in from a high pressure area, becoming warm and taking up moisture. As the air rises, it cools and clouds and rain are formed.

In the high pressure area, air moves at low levels towards the low pressure area. That air is replaced by air coming from higher levels. The sinking air warms, relative humidity falls, and the clouds disappear. Strong sunshine leads to cloud formation on the periphery of the cell of sinking air (fair weather cumulus).

1. Sinking air over the cooled land;
2. at ground levels, outward-moving air that lifts after warming above the water surface;
3. at heights, inward-moving air originating over the warm water surface;
4. at ground level, outward-moving air again warms and rises over the water surface and, as it rises, moves in and increases the pressure again over the land.

In both cases, it is the temperature difference between the air masses leading to pressure differences that begins the movement. The presence of a pressure gradient shows that forces are at work on the air and air movement begins – that is to say, a wind blows.

The relationship between air masses of different temperatures can be studied with the help of a small experiment. All that is needed are a candle and two inter-communicating rooms – one heated, the other not. If the door between the rooms is opened and a lighted candle placed on the sill, the flickering of the candle will serve as an indicator of an exchange of air. It is the cold air from the unheated room that flows into the heated room at or near floor level. The cold air pushes under the warm air. This lifts to higher levels and moves into the room that began with cold air. Gradually it sinks and takes the place of the departed cold air. The candle flame indicates the direction of flow. Movement comes to an end once the cold air occupies the lower levels in both rooms, and the warm air the upper levels. In an ideal case both air masses are

Air movement near the ground is spiral, as a result of being deflected by the Coriolis force (see p. 175). The air moves around the centres of high and low pressure areas in circular paths. In the northern hemisphere, the Coriolis force deflects the movement to the right around highs and to the left around lows. In the lowest levels of the atmosphere, less than 2000 metres (6500ft), the relief in the earth's surface has a braking effect. This is how the exchange between low and high pressures takes place: air moves from the high to the low.

separated by a horizontal boundary surface. What are the air pressures? While the exchange is taking place, the pressure at floor level in the room with cold air is higher since the air is heavier. The pressure is lower under the warm air. The movement follows the pressure gradient, the cold air flowing to the warm at floor level. When movement has finished, the pressure at floor level is considerably greater than in the warm sector above.

Although this is only a model (in the free atmosphere there is, obviously, nothing which serves as room walls), it does draw attention to an important phenomenon – circulation with air movement at ground level (movement with high pressure) and air movement at higher levels (movement with low pressure). The direction of movement at higher levels is the opposite to that at lower levels.

1. At ground level cold air moves to warm.
2. At higher levels warm air moves to cold.

This is true for circulations caused by temperature differences. From observations of the weather it is clear that air does not always move from cold to warm near the ground.

Effect of the earth's rotation on the weather

The patterns in which air moves would be much simpler if the earth did not rotate. Differences in pressure would be immediately removed by air moving from high to low. The effects of rotation are, however, interposed, with the earth's centrifugal force, creating a deflection of the moving air. It was the French physicist, C. G. de Coriolis (1792–1843), who first showed this deflection to be an apparent acceleration, turning a moving object from its path, if the movement of the object is linked by inertia to the circular movement of the earth ('Coriolis acceleration', 'Coriolis force').

The rotation of the earth has effects in both the vertical and the horizontal planes. A body moving eastwards in relation to the earth must be moving more rapidly than the earth around its axis. The body's centrifugal force must therefore be greater than if it remained at rest, and the gravitational attraction of the earth less. The eastwards moving body is therefore apparently lighter, the westward heavier. This vertical component of the Coriolis acceleration is not apparent at low speeds. It has an important effect on projectiles, aircraft and space vehicles and in the troposphere, with winds at normal speeds, it plays a subsidiary role.

The horizontal component of the Coriolis acceleration becomes more important in the higher latitudes, its effect being non-existent at the equator. The horizontal component deflects an airstream moving from west to east:

> to the right in the northern hemisphere
> to the left in the southern hemisphere

The deflection does not affect only air movements parallel to latitude. The horizontal component of the Coriolis acceleration affects all air movement, no matter what their direction, for example, from north to south. The result is that there always is a deflection to the right in the northern hemisphere and to the left in the southern hemisphere. The only exception is an air movement along the equator.

The Coriolis acceleration so far discussed, both vertical and horizontal, pertains to air movement in a straight line; however, air movements are usually curved and in this case the resulting centrifugal force is not only caused by the rotation of the earth but also depends on the curvature of the air movements and on their speed.

There is also a third force at work on moving air. This is the friction resulting from movement over the earth's surface. The land's relief is particularly sharp in this respect – more friction is therefore caused by air movement over land than over sea, with energy being expended to overcome the friction and creating a braking effect.

The influence of the earth's rotation on extensive air movement is very great. As Prof. Heinz Renter says in *The Science of Weather*: 'Without an understanding of the effect of the Coriolis force, the air movements resulting from pressure distribution that we can distinguish in the daily weather map, are completely inexplicable.'

On a weather map, the distance between the isobars (see p. 109) indicates the strength of the pressure gradient, and hence windspeed, with a force directed from high pressure to low. The Coriolis force then has the effect of diverting the force of the pressure. With increased windspeed, the diversion becomes more marked. When a balance is achieved between the Coriolis acceleration and the pressure gradient, a near-stationary condition is achieved.

What then is the relationship between air movements and high or low pressure systems? The friction effects that brake air movements result in a low pressure area in a spiral air movement. In the northern hemisphere the spirals have a curvature to the left, in the southern hemisphere a curvature to the right. In low pressure areas the air movement is upward, into levels of less pressure. It is therefore accompanied by cooling until 100 per cent relative humidity (see p. 58) is achieved, clouds are formed (by condensation of water vapour) and in some circumstances rainfall. The centrifugal force, operating in the spiral movement, is by nature exerted outwards. All the various forces, acting together, produce an air vortex – the circular movement around the centre of the low pressure. Such vortices, with vertical axes, are particularly strong in hurricanes. It is remarkable that, in the central point of a hurricane, hardly any wind is noticeable (eye of the storm).

From a high pressure system the air spirals outwards, with a curvature to the right in the northern hemisphere and to the left in the southern hemisphere. In a high pressure area the air movement is downward. The air comes under higher pressure and is warmed. Since the air can take up more water vapour, any clouds are absorbed and the sky clears.

On the weather map the isobars are circular or elliptical for both low and high pressure areas (see p. 98). The pattern of isobars indicates the direction of air movement.

Air movement in the northern hemisphere has higher pressure to the right, lower pressure to the left.
Air movement in the southern hemisphere has higher pressure to the left, lower pressure to the right.

If the isobars are close together, the pressure gradient is strong and winds are high. Because there is less friction, speeds are greater over the sea and at great altitudes.

In European conditions wind strengths are approximately as follows:

distance between 5mb isobars	wind strength
600km (373mi)	light breeze (Beaufort Scale 2)
500km (310mi)	moderate breeze (Beaufort Scale 4)
400km (248mi)	fresh breeze (Beaufort Scale 5)
300km (186mi)	strong breeze (Beaufort Scale 6)
200km (124mi)	near gale (Beaufort Scale 7)
100km (62mi)	strong gale (Beaufort Scale 9)

From the above it is clear that the distribution of pressure is very important to our understanding of weather, but it would not be accurate to suggest that weather forecasts can be based solely on an assessment of pressure systems.

The deflecting function of mountains

The surface of land influences not only the frictional process and the resulting loss of energy – mountains and mountain ranges have a direct effect on air movements near the surface, forcing them to be deflected or change direction. This characteristic is, in turn, linked to regional differences in phenomena. The most important geographic barriers in Europe are:

1. The Norwegian mountains, orientated north-northeast to south-southwest, and rising to 2500 metres (8200ft) in height.
2. In central Europe the Alps – the highest mountains in Europe – extending from southern France to Austria. Their total length is 1200km (750mi), the average height 1400 metres (4600ft), and the highest, Montblanc, is 4807 metres (15,781ft).
3. In Spain the Pyrenees, which are 435km (270mi) long; Pico de Aneto is the highest at 3404 metres (11,168ft). Its western extensions are in the Cantabrians, which are up to 2648 metres (8687ft).

In addition to these large mountain ranges, there are other hills and mountains of more local significance such as the Welsh Mountains, the Highlands in Scotland, the central mountains of Germany, such as the Black Forest, the Cevennes and the Vosges in France, the Dalmatian and Albanian coastal ranges, and the Apennines in Italy.

Diverse relief and an extremely indented coastline are also characteristic of Europe, there being numerous islands and peninsulas. The largest peninsulas are:

Fennoscandinavia (Norway, Sweden, Finland)	1,240,000km² (479,000sq mi)
Iberia (Spain and Portugal)	584,000km² (225,000sq mi)
The Balkans	468,000km² (181,000sq mi)
Italy	149,000km² (58,000sq mi)

There are, too, a variety of coastal forms: fiords in Norway, skerry guard coasts of Scandinavia, loch-indented Scottish coasts, chalk-lined cliffs in southern England, the mud flats of eastern Friesland and Holland, longitudinal coasts with parallel flanking mountains as in Yugoslavia, great estuaries as of the Elbe, Weser, Thames and Loire, and deltas such as those of the Rhône, Rhine, Vistula, Danube and Po.

This variety of form brings a great interspersing of water and land surfaces, with a resulting great variety of local weather situations. In mountains, wind movements are blocked, and fall winds occur (see p. 183), both having important effects on the weather. This is also true of the more regular mountain and valley winds (see p. 32). There are many spatial and temporal contrasts in humidity. Mountains and hills lift the air masses in movement towards and across them, with resultant cooling and, depending on the humidity of the air, cloud and rain. Lower and isolated hills tend to be encircled by moving air. Valleys can funnel winds in certain directions. If the wind direction and the orientation of the valley coincide this can lead to a damming or constricting effect. An example of this is the mistral (see p. 182) in the Rhône valley. In such narrow features the constriction increases the windspeeds, and at the exit of the valley the wind spills out in a semi-circle.

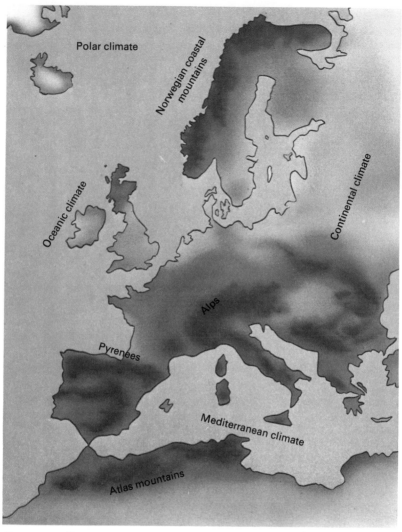

Polar climate

Norwegian coastal mountains

Continental climate

Oceanic climate

Alps

Pyrenees

Mediterranean climate

Atlas mountains

European climate

Geographical latitude, altitude above sea level, distance from the sea, influence of mountains, state of the ground, vegetation and average atmospheric conditions are essential factors in determining the climate of any given region. Climate emphasizes certain basic features of the weather: Atlantic (oceanic) climate – moist and cool, no cold winters (snow rare) and no hot summers; Continental climate – dry and extreme, very cold winter, hot dry summers; Mediterranean climate – relatively dry summer, winter rainy (snow infrequent) and mild.

Wind systems

Wind movements are one of the major variations in weather, contributing to temperature change, cloud formation and rainfall. The winds of the earth show a wide range of characteristics. European winds are characterized by a relatively great variety in both space and time, but the wind systems on either side of the equator – the trades – have an astounding degree of constancy. This was of great importance in the days of sail. The trades belong to the so-called 'planetary' wind system, the system owing its origin to energy from solar radiation. The trades dominate in low latitudes (see p. 116).

The northeast trades and the southeast trades come together at the equator, the belt where they meet being described as a 'convergence zone' and characterized by rising air streams, strong cloud formation and heavy rain. Because of the lack of wind within the convergence zone, sailors knew it as the 'doldrums'.

North of the trades, in both the northern and the southern hemispheres, there is a subtropical high pressure zone between 30° and 35°, known to sailors as the 'horse' latitudes. Good weather predominates here, winds are weak and rainfall small. The zone possesses great pressure stability, and is also part of the planetary wind system. An area of high pressure in the north Atlantic centred south or west of the Azores belongs to the subtropical high pressure zone. Known as the Azores high, it is frequently mentioned in weather reports and plays an important part in European weather. On the north side of this high pressure area, marine air flows towards Europe. In the summer months, the Azores high reaches as far as Spain and the south of France, in winter, it is generally about 30°N. Its movements are particularly marked in spring and autumn. The high does not only bring good weather – indeed, it frequently helps direct westerly air streams towards Europe (see p. 169).

The planetary wind system, taking in the convergence zone at the equator, the trades and the subtropical high pressure zone, changes position in accord with the sun's movements – north in summer, south in winter (with a maximum change in movement of about 10°). This seasonal movement creates phases lasting several months each. The northerly latitudes are reached in March, the northernmost in midsummer, southerly latitudes are reached in September, the southernmost in midwinter. This movement is of great significance to the weather of the tropics, determining the coming of the rainy season and the formation of tropical cyclones.

Another wind system covering a large area and having a seasonal recurrence is known by the collective term 'monsoon'. This system is known as 'terrestrial', since it originates out of the seasonal differences in the heating of land and sea. The mechanics of this movement are the same as those in land and sea breezes (see p. 32), but alternating between summer and winter instead of between day and night. Monsoon winds are of greatest importance in southern and southeastern Asia. From the cooler sea, air masses move in summer to the warmer continent, bringing heavy rain (the summer monsoon). In winter, cooler air from the continent moves out to the warmer sea, bringing in the winter monsoon, drier weather with little cloud. The occurrence of a similar phenomenon in Europe is dealt with on p. 32.

In the middle latitudes, such as those of Europe, a continuous exchange of air from the subtropics and the polar regions is typical and the air streams are usually from west to east. The polar low in the northern hemisphere is situated at about 60° north – somewhere in the line Scotland, south Norway, south Sweden and the Gulf of Finland.

It is relatively narrow but many extra-tropical depressions move towards it (see p. 195).

Air movements of the earth can be described simply in terms of the differential heating of land and sea, and the observed temperature gradient from equator to pole. A compensating flow in the upper atmosphere, balancing the ground level movement, can also be demonstrated. The sole exceptions occur in the middle latitudes, with west winds at higher altitudes as well as at ground level. An explanation is dependent on research into the upper layers of the troposphere, in which the so-called high wind belts were discovered, with jet streams having speeds of hundreds of kilometres an hour. Above the subtropical high pressure zone, one such air stream was discovered moving from west to east.

The jet streams are an expression of great turbulence in the troposphere, the result of the combined effects of the various forces that set air streams in motion, accelerate and brake them (see p. 28).

Fall and vortex winds

Winds which are restricted to identifiable regions may be classified by a link with local geographical features – a mountain chain or a desert and these contrast sharply with the larger wind systems.

Land and sea breezes have already been identified (see p. 32), as have the winds associated with storms (see p. 84). Then there are fall winds worldwide wherever mountains block or deflect air movements – the best-known of which is the Föhn wind (see p. 46). Generally the Föhn blows from south to north, but there is also a north Föhn on the south side of the Alps, bringing cold air to the Mediterranean.

A typical fall wind in the Adriatic Sea is the *bora*, a dry and relatively cold wind blowing from the barren slopes of the Dalmatian and Albanian coastal mountains. It is a squally, stormy wind, important to yachting in the Adriatic. The bora develops under various conditions, usually in winter with a high over the Balkans and eastern Europe. The associated weather is fine, dry and cold. The stimulus comes from a difference in pressure between the coastal mountains and the warm sea. Because of the stability of the pressure systems in winter, the wind may blow for a few weeks. There is also a 'cyclonic bora', which has its origins in a deep depression over the Adriatic coupled with a high over the Balkans. A cyclonic bora lasts only a day or even a few hours but brings storms and sometimes heavy rain. The bora is a fall wind but, nevertheless, is cold. The air in the mountains is cold and, despite warming as it descends, it still contrasts with the warm air over the Adriatic. A typical warning of the advent of a bora is the build-up of cloud on the summits of the coastal mountains.

The *mistral* is the fall wind native to the south coast of France, a cold dry wind in the Rhône delta between Avignon and Marseilles. The causes are a deep depression over the warm sea and high pressure associated with cold dry air over the Cevennes and the French Alps. The depression sucks in air from the land, and the constriction of the Rhône valley increases the speed.

The *sirocco* is a fall wind off the north African coast, a hot dry southerly wind. It develops with a depression over the Mediterranean, drawing in air from the Atlas mountains. The sirocco brings dust and sand from the Sahara, and its effects occasionally reach central Europe and, on very rare occasions, even southern England. In

Many wind systems depend on unstable air stratification, the prerequisite for thermally-determined air streams.

sea wind

land wind (sea coast)

upslope wind,

downslope wind   Föhn wind

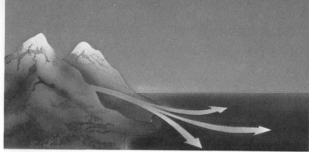

Mistral

Solar radiation (warming the air from the earth's surface) or the arrival of cold air in the upper layers of the atmosphere leads to unstable conditions.

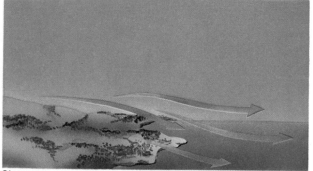

Bora

Sirocco

Mountain wind

Valley wind

sea wind (mountain)

rain shower

storm wind

whirlwind

the western Mediterranean the sirocco produces squally wind vortices. There is a wind in the eastern Mediterranean also known as sirocco, but it is not a true fall wind. Its origins lie in a deep depression coming from the west, producing a turbulent air stream.

The boundary zones between warm and cold air masses are the most favourable locations for the development of vortices, particularly in the warm season. Such vortices range from small dust devils and water spouts to tornados and tropical hurricanes possessing tremendous destructive effects. Whilst the tropical hurricanes and typhoons are a type of cyclone or depression with a diameter of 100 to 600 kilometres, tornados are only a few kilometres across. They owe their violence to the wake of a low pressure system.

Apart from dust devils and water spouts, violent revolving storms are rare but not impossible in Europe. In Britain a wind storm of this type did serious damage in Stevenage, the new town in Hertfordshire.

Tornados are less rare in Europe than is generally thought. Although they lack the drama of the hurricanes and tornados of America, where their regularity makes them a constant threat to people's livelihoods, it is now thought by scientists that a tornado can be expected every five to ten years. They can occur anywhere in Europe and, over the last decade, there have been reports of their occurrence in France, Belgium, Hungary, the Netherlands, Poland, Switzerland and the USSR.

Many regional winds have names that are used in the media and in weather reports. The following is a selection:

name	region	type
Baguio	Philippines	tropical revolving storm
Bise	Switzerland, France	with high pressure in the Alpine foreland, from north to east
Blizzard	North America	Snow storm with cold air inbreak
Bora	Yugoslavia	cold fall wind
Khamsin	Egypt	hot desert wind from southeast
Chinook	US	warm fall wind
Cordonazo	Central America	tropical revolving storm
Etesian	eastern Mediterranean	regular dry summer winds from north to northeast.
Föhn	Alps	warm fall wind
Harmattan	coast of western Africa	dry dusty northwind (part of the trades winds)
Hurricane	West Indies	tropical revolving storm
Mauritius hurricane	Indian Ocean	tropical revolving storm
Mistral	southern France	cold fall wind
Monsoon	southern Asia	terrestrial wind system
Norther	North and Central America	storm as the result of a cold inbreak from the north
Pampero	South America	storm as the result of a cold air inbreak from the south

Trades	tropical latitudes	planetary wind system
Samum	northern Africa	sandstorm
Suestados	Argentina	storm from southeast
Sumatras	Malacca Straits	fall wind
South Seas hurricane	southern Pacific	tropical revolving storm
Typhoon	northern Pacific	tropical revolving storm
Tornado (1) ('twister')	North America	large whirlwind
Tornado (2)	western Africa	thunderstorm
Whirlwind	all parts of the earth	local revolving storm
White squalls	West Indies	fall wind
Williwaws	Argentina	fall wind
Willy-willies	Australia	tropical revolving storm
Zonda	Andes	fall wind
Cyclone	Indian Ocean	tropical revolving storm

Air masses and fronts

The weather in European latitudes is dominated by an exchange process between cold and warm air. Under the same pressure conditions cold air is heavier than warm and, therefore, always predominates ground level. Warm air spreads out at greater altitudes. It is important to note that 0°C at 4000 metres (13,100ft) is more difficult to assess, as an indication of an air mass, than at ground level – yet, the meteorologist must evaluate the temperatures of every air mass of significance in weather forecasting. *The boundaries between cold air masses and warm air masses are where most important phenomena in weather events occur.* If cold air is moving forward, then the mass is called a 'cold' front, if warm air is advancing, then it is a 'warm' front.

The temperature of an air mass is dependent, by and large, on the latitudes from which it comes. There are particular regions which act as sources for the air masses that contribute to European weather, such as the polar region and the horse latitudes. The following is an outline of the more significant air masses, their source and their characteristics:

name	source	characteristic
Polar air (arctic air)	polar region	cold, dry
Maritime polar air	polar region and a long route over the ocean	cool, moist, particularly in the lower layers no longer quite so cold
Continental polar air	polar region and a long route over continental Europe	cold and very dry or even extremely dry
Tropical air	subtropical high pressure zone	warm dry
Maritime tropical air	Atlantic, Azores region	warm, moist to very moist
Continental tropical air	northern Africa, Asia Minor	warm and very dry

The horizontal movement of cold and warm air masses over relatively long routes, their meeting and mixing is very important in the development of all weather. These mixing processes take place in the troposphere – the lowest layers of the atmosphere.

The warm and cold fronts have already been discussed (see p. 48). As a cold front advances, warm air is lifted and, as a warm front advances, again the warm air moves up over cold air – so that both phenomena are associated with the formation of cloud and, in most cases, with precipitation.

The weather before and during the passage of a warm front is relatively calm, but the rain belt can be wide. After the passage of the front, air temperatures begin to rise and the clouds break. Generally it is found that improvement in the weather is sharper after the passage of a warm front than after the passage of a cold front. The change in pressure after the front has passed may be very slow or a further fall may even be registered.

The weather before and during the passage of a cold front is stormy in the true sense of the word. Even during the warmer periods of the year the cold air inbreak is accompanied by snow showers or hail and sleet. Frontal thunderstorms often occur. Although the cloud and rain belt is relatively narrow and the increase in temperature after the front deceptively rapid, it will take some time before the weather becomes settled. The pressure will normally increase and the rise may be very sharp.

During a warm front

	before the front	*at the front*	*after the front*
Clouds	appearance of cirrus, followed by altostatus and altocumulus	deep rain clouds, nimbostratus	clouds break
Pressure	falling	falling	constant, or still falling
Wind	southeast to south, fresh and cool	south freshening	southwest to west, fresh and warm
Temperature	sinking	rising	rising
Visibility	deteriorating	bad	increasingly good
Weather	grey skies drizzle	overcast, heavy rain	some lightening in the sky, rain stopping

During a cold front

	before the front	*at the front*	*behind the front*
Cloud	cirrus at heights, massive roll of cumulonimbus	rain clouds – nimbostratus, cumulus	clearing, often very quickly
Pressure	falling	rising or, in some circumstances, rising strongly	rising
Wind	southwest, strong and cool	west to northwest, stormy and cold; also squally	northwest strong and sometimes cool
Temperature	falling	falling	falling
Visibility	improving	bad	good
Weather	driving cloud, gathering in west and northwest	sky covered with quickly-moving cloud and cloud trails under certain circumstances thunderstorms	tendency to rain showers – remains changeable weather

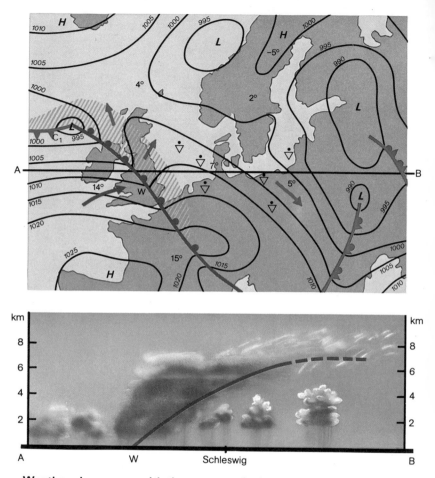

Weather phenomena with the passage of a front – 1

Weather map of 11 November, 1947 (above) and vertical cloud section along line A–B (below), as observed at Schleswig, northern West Germany. Weather behind a front with showers, following a depression over western USSR. Another depression is situated over Ireland, and warm air moving northwards through the western Mediterranean is pressing into central Europe on its south side. This warm air moves in above the cold air. Preceding the warm front (W) is a rain belt (shaded green on weather map). Strong cumulus development is indicative of the weather behind the cold front; high and middle-height stratiform clouds herald the coming warm front.

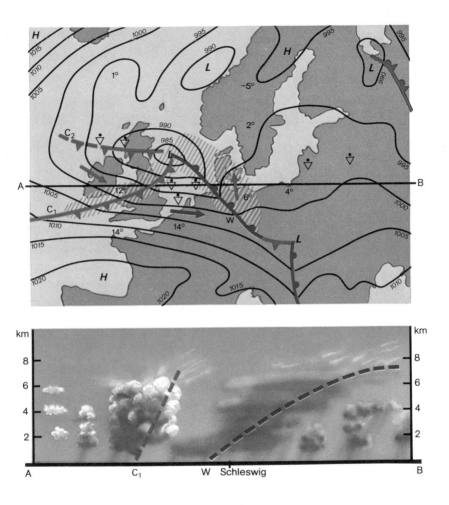

Weather phenomena with the passage of a depression – 2

Weather map of 12 November 1947 (above) and vertical cloud section along line A–B (below), as observed at Schleswig, northern West Germany. The depression over Ireland has now reached the North Sea. The warm front (W) lies on the west coast of Schleswig-Holstein. The cold front (C) is over the western part of the North Sea and is pushing quickly towards the warm front. Thick cloud and rain are over Schleswig. The proximity of the cold front hinders the break-up of the cloud cover, and brings showers. After the cold front has passed, the clouds slowly break up.

189

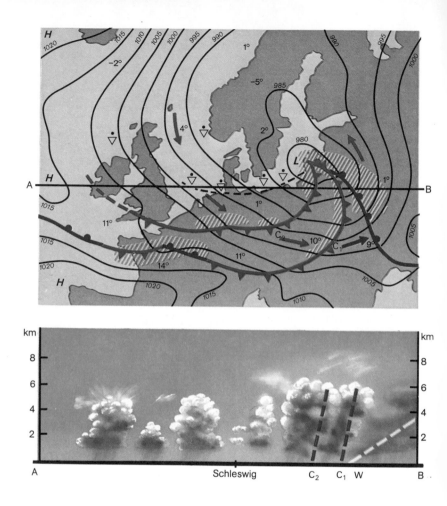

Weather phenomena with the passage of a depression – 3

Weather map of 13 November 1947 (above) and vertical cloud section along lines A–B (below), as observed at Schleswig, northern West Germany. The depression moving from Ireland via the North Sea and Schleswig-Holstein has now reached the Baltic Sea and western USSR. Cold air reaching great heights has now covered all of Germany, bringing showery weather to the northern German coasts. Cumulus clouds and blue skies are present with good visibility away from the shower areas themselves. (W = warm front, C_1, C_2 = cold fronts.)

The struggle between warm air and cold air is particularly impressive in a depression. In European latitudes, cyclonic activity often covers a wide area and involves at least two air masses of different temperatures. Although they are more frequent in autumn and spring, they can occur during any season of the year.

In middle latitudes of Europe, the eastward-flowing warm air cannot move past the westward-moving cold air without disturbance. Mountains can have the same effect alone in damming and deflecting air masses, as for example the Canadian Rockies, the mountains of Greenland and Norway, and the Urals in the USSR. The deflection of cold air movement towards the south leads to large-scale inbreaks into the warm air. This leads in turn to the formation of vortices, made clear in cyclonic activity by a spiral pattern of clouds – as can be clearly observed in satellite photographs (see p. 9). In depressions, the fronts occur normally, as described, and generally in sequence. In its initial stages, a cyclone affects only the lower layers of air, extending subsequently into the upper layers.

Cold fronts and warm fronts often show departures from the norm. There are cold fronts that at first show the cloud sequence associated with warm fronts, and only gradually is the normal cold front pattern established. In contrast, in summer the warm air moving forward may be much more unstable so that cumulus cloud will occur with the stratus clouds – and occasionally rain from the warm front will be accompanied by thunderstorms. Furthermore, an intermediate high pressure system can change the appearance of a warm front so that it is characterized only by a group of cirrus clouds.

Stages in the life of a cyclone. It is now important to examine the stages of a cyclone, as it develops as a result of contact between warm and cold fronts:

1. Two inbreaks of cold air enclose a zone of warm air
2. The warm air glides up over the eastern cold air inbreak (warm front)
3. The western cold air pushes into the warm section (cold front)
4. The warm air sector is continuously compressed
5. The western cold front overtakes the warm front
6. The remnants of the warm sector are lifted to higher levels the cyclone and is 'occluded'.

The depression of which the cyclone is part moves from west to east. The time involved in the stages outlined above is about 48 hours, but both structures and speeds are extremely variable. By the time that the cyclone is occluded, movement has almost ceased (stationary depression).

The weather in this cyclone corresponds generally with that described as at the warm front and at the cold front (see p. 189). The occlusion is 'frontal' in character (two cold air masses of differing temperature meet).

At the occluded front

	before the front	*at the front*	*after the front*
Cloud	rain cloud, nimbostratus	overcast, nimbostratus and cumulus	clouds break
Pressure	falling	falling	rising
Wind	freshening, northwest	increasing, northwest	dying down, cool, northeast
Temperature	falling	falling or remaining cool	falling
Visibility	moderate	bad	good
Weather	overcast and rain	overcast and rain	break of clouds, isolated showers

Two particular phenomena make cyclonic weather conditions even more remarkable:

1. The trough. This is characterized by very low pressure after the passage of the cold front of a cyclone, and is a deepening of the depression. In front of the trough, the winds behind the cold front die away, but, as the trough arrives, winds gradually rise to storm strength.

   A further variation is the cold low (or cold pool), usually at high levels in the stratosphere. Extraordinarily heavy rain occurs in these conditions. For example, in central Lake Zürich at Horgen it rained in August 1975 without a break for 42 hours. During this period 79 litres fell every square metre (17.45 gallons per square yard). The rain, which began suddenly, lasted 42 hours and fell without variation in intensity. This was the result of a small inclusion of cold air in the region of Genoa, which led to heavy rain on both sides of the Alps.
2. The intermediate high. This is a small wedge of high pressure between the cyclones bringing a transitory improvement in the weather.

In Europe depressions occur usually in a series, one after another, as a sequence of disturbances along the boundary between warm air and cold air ('polar front'). In the movement from west to east the depression takes a well-defined route, usually following the sea because of the lack of mountains as obstacles to movement. The land, and mountainous terrain particularly, accelerates the rate of speed at which an occlusion develops.

Depressions move from west to east changing in form and intensity during the course of their journey. The uppermost figure shows a typical depression by the time it has reached central Europe. The more rapidly-moving cold air has compressed the warm sector, lifting the warm air from the ground into the occluded sector as far as P, the occluded front – shown as an alternation of the cold front and warm front symbols. In the occlusion the two air masses, originally separated by the warm air, have met. The middle figure shows a section through a cold occlusion, very cold air being present in the following air mass and bringing changeable weather with showers. In the lowest figure a warm occlusion is shown in section – the contrast between cold air and warm air is less marked and the lift of the warm air is completed in 'calm' conditions with the formation of stratus cloud. Steady rain predominates. With occlusion the effectiveness of a depression comes to an end. The struggle between cold air and warm air has ended. The passage of clouds marks the event for the observer.

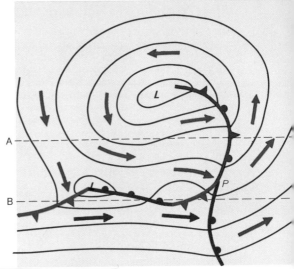

A = cold air sector
B = warm air sector
blue arrows = cold air
L = depression
red arrows = warm air
C = cold air
Ca = particularly cold air
W = warm air

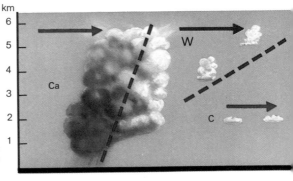

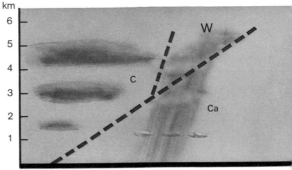

Influence of the weather on mankind

The tendency for various weather phenomena to occur at particular seasons has led to the passing down from one generation to the next of proverbs and weather lore. Examples are the spring inbreaks of cold air, giving rise to rain ('April showers'), the period of late spring frosts ('Blackthorn winter'), and the fine settled conditions associated with a high pressure in late summer ('Indian summer'). The behaviour of animals and plants may also be indicative of changes in the weather. For example, with the coming of a storm, bees return to their hives. Many animals that prey on insects (birds, frogs, fish) react to the height at which their victims (flies, gnats, midges, etc.) are active. Insects swarm in hot dry weather at greater heights than in cool wet weather. In this, the electrical state of the atmosphere plays an important role – as it does in man's reaction to weather.

At a medical congress in Berlin in 1975, it was stated that every third adult between 20 and 50 years of age could sense or have an adverse reaction to certain weather changes – generally more prevalent in women than in men. The discussion at the congress included a wide range of topics, such as circulation and heart troubles during a heat wave, problems of European sportsmen performing in countries with higher air humidities, and the increased effects of alcohol during a change in the weather. One medical report advised that persons with a tendency to an adverse reaction to weather changes should take more salt, sugar and water. At critical weather periods, such as during Föhn winds, such persons should drink their tea and coffee sweeter, eat their food more highly salted and drink more. Be that as it may, one can assert without contradiction that the whole field of the affect of weather on human health requires much further research.

Storms do not necessarily bring only damage and death in their trail. For example the storms of early 1976, which were a catastrophe for part of western and central Europe, forced large volumes of North Sea water, with a higher salt and oxygen content, into the Baltic through the Danish straits. The oxygen content of the Baltic had been declining for some 25 years, to such an extent that the fish population and the local flora were threatened. The heavier North Sea water has produced a constituent change in the Baltic water, sinking to the lower levels and re-oxygenating the water mass, bringing some positive and beneficial effects!

Typical trajectories of cyclones over Europe

1. The cyclone moves southwest–northeast from west of the British Isles. The cyclone affects weather only on the north of the mainland.
2. From north of the British Isles, the cyclone moves west–east over the North Sea and then Scandinavia.
3. From west of the British Isles, the cyclone moves across the northern parts of the British Isles, then west–east across the North Sea and over southern Scandinavia and the Baltic lands.
4. From the southwest, the cyclone moves northeast passing between England and France, over the North Sea and above Denmark to the Baltic region.
5. From the west, the cyclone moves across France to the western Mediterranean and, then, either over the southern and eastern edge of the Alps to east Europe or, more southerly, over the Balkans.

Meteorological services and weather forecasting

Everyone is interested in weather forecasts, and for some people – farmers, seamen, fishermen, builders, etc. – they are vital to their livelihoods. The task of a meteorological service, and the meteorologist employed in that service, is in no way limited to the daily forecast. As in other sciences, there are specialized fields with responsibility for increasing our knowledge of weather and climate – and the scientific inter-relationships with other disciplines. Examples of these specialized fields are agricultural and medical meteorology (together biometeorology), aviation meteorology, marine meteorology, radio meteorology, radar meteorology and satellite meteorology.

The term meteorology (the study of processes in the atmosphere) was coined by the Greek philosopher Aristotle (384–322 BC). However, men have always observed the weather, even if, for millennia, that observance was manifested in the propitiating of the gods of the elements (sun, rain, thunder, etc.). Long ago, regional knowledge of phenomena was gathered together, giving rise to the various proverbs and sayings that constitute the weather lore. Descriptions of the weather over a period of years, and of associated weather lore, have been published in almanacs and annuals for centuries. An early example is the 'True century-calendar', based on the observations of Abbot Mauritius Knauer (1612–64), and later revisions of his work. Knauer named his work *Calendarium oeconomcium practicum perpetuum*. It contained astronomical observations as well as weather observations, and an explanation of weather in which astrological influences could be recognized. Knauer was much attracted to long-term weather observations. In the foreword to his calendar, he offered comfort when forecasts were strikingly wrong, pointing out that everything cannot be known and that God cannot be prescribed in scale or aim. A list of severe winters, droughts, floods etc was given by Dr Thomas Short in his work 'A General Chronological History of Air, Weather, Seasons, Meteors etc.' 1749.

An early English example of careful observations and recording of the weather is the *Gardener's Calendar* of Gilbert White (1720–93). This now famous naturalist recorded weather observations in his journal for many years (see p. 232).

Verse proverbs about the weather go back to Babylonian times. Various English proverbs, and an explanation of their original meanings, are given in *Seasonable Weather* by Lionel P. Smith (1968). Running through proverbs on weather is their identifying of conditions with geographical features. Weather conditions next to mountains were seen to be quite different from those in the lowlands, the times of ripening and harvesting and the times of sowing quite different in the uplands from those of the more fertile and climatically-favoured lowlands. Thus, the proverbs assembled in one locality could not be deemed applicable to others. If, today, the original time and place of the proverb is no longer known, then the knowledge contained within that proverb is in all probability lost.

This applies particularly to the various recurring warm and cold spells such as Blackthorn winter and the Ice Saints. The Scottish meteorologist Buchan, in 1869, using observations and various Scottish weather sayings, distinguished six cold periods:

    7 – 14 February
    11 – 14 April
    9 – 14 May
    29 June – 4 July
    6 – 11 August
    6 – 13 November

and three warm periods

    12 – 15 July
    12 – 15 August
    3 – 14 December

In the 1920's and 1930's these Buchan cold and warm spells became widely known and were taken to apply throughout the British Isles although Buchan had worked only with Scottish data. Examination of the temperature at Kew for the period 1871–1929 led to the conclusion that none of the Buchan cold and warm spells applied there.

Indeed it can be seen that although the cold spell 9–14 May does overlap with the Ice Saints period in central Europe (12–15 May) the warm period associated with St Luke's Day (18th October), the 'Golden October' of German speaking lands, is not a Buchan warm spell.

A number of weather proverbs or days considered to have special meteorological significance has also been tested statistically. An example is St Swithin's Day. According to the legend St Swithin (canonised by popular usage and not by the church) was buried first outside the Old Minister in Winchester, in compliance with his wish that the rain should fall on his tomb. On 15 July 971 his body was transferred to the New Minster (both Old and New Minsters have vanished but were closer to the site of the present cathedral), and since his wishes had been ignored heavy rain fell, continuing for 40 days. The legend is charming but an analysis of the rainfall statistics fails to provide any support. A rain day is a day with at least 0.01″ of rain. In the period 1933–1962 at Kew a wet St Swithin's Day was followed on average by 14.3 rain days. There were an average 15.4 rain days after a dry St Swithin's Day. That is to say there was little real difference during the following 40 days whether the 15th July was wet or dry.

With both the Buchan cold and warm spells, and with days such as St Swithin's, thought to have a special meteorological significance, there is a further difficulty. On 2nd September 1752 the old Julian calendar was replaced by the Gregorian calendar, as calendar time had become progressively behind astronomical time. The day following the 2nd September 1752 became the 14th September and eleven days were dropped. Obviously this provides a serious complication since the significant periods or days have changed their position with respect to the passage of the seasons.

The newsheets of the 17th and 18th centuries giving details of comets are of particular historical interest. It was then believed, as it had been in ancient and medieval times, that comets were part of the earth, related in some way to the clouds. It is clear from these newsheets that the atmospheric phenomena had been very carefully observed.

The beginnings of scientific weather observation were founded in the development of the thermometer and the barometer. With the help of individual observations, our understanding of the weather began to advance. Regular observations began in the 18th century. The oldest mountain weather station in the world is Hohen Peissenburg, at 988 metres (3241ft), in Upper Bavaria. Systematic observations have been made there since 1781, and today the German meteorological service maintains an observatory there for radar meteorology and rainfall physics.

Using barometer and thermometer readings the German physicist Heinrich Wilhelm Brand (1777–1834) produced the first synoptic weather map in 1820. Evaluation of weather reports was hindered by the lack of any means to convey information quickly, but this was resolved with the discovery of the telegraph (Samuel Morse in 1837) and the telephone (Alexander Graham Bell in 1876).

A telegraphic weather report was first published in England, in 1848 in the *Daily News*, and at the Great Exhibition in London 1851 a daily weather map was produced, with weather reports from all over Europe, the reports coming from 22 weather stations. It was probably the destruction of a French fleet by a storm during the Crimean War that led to the demand for a daily weather map in France (from 1855), and focused attention on the importance of regular weather forecasts for shipping. In this century there has been the parallel need to have established weather services for aviation. Aviation meteorological services were formed during the First World War.

International co-operation for the exchange of meteorological information began very early. At the international congress of meteorologists in Vienna, held in 1873, the International Meteorological Organization (IMO) was founded. Its tasks were taken over in 1951 by the World Meteorological Organization (WMO), a specialist branch of the United Nations with offices in Geneva. Virtually all countries belong to the WMO, the main tasks of which remain the improvement of the net of observation stations around the world, the establishment of a more uniform reporting system and the speeding up of the rate of transmission of information. The eight technical commissions of the organization deal with specialist areas, and act in parallel to the divisions in national meteorological services. These are:

1. Synoptic meteorology, for the preparation and publication of weather forecasts, ready for their dissemination by radio, television and newspapers.
2. Aviation meteorology, for advice on flight conditions on air routes, and the establishment of stations to facilitate the gathering of this information.
3. Climatology, for observation and evaluation of climatic elements that are of growing importance in public life (e.g. agriculture, pollution, planning, public health).
4. Physics of the atmosphere, dealing with research work on the atmosphere and the continuous observation of some of its elements (e.g. radioactivity, measurement of high altitude winds).
5. Agricultural meteorology, with the aim of building a knowledge of the climatic elements valuable in agriculture; a special field is forest meteorology.
6. Hydrometeorology, dealing with the formation of precipitation and the water balance in nature.
7. Maritime meteorology for the investigation of weather and climate at sea.
8. Instruments and methods of observation. This is concerned with the further development of meteorological instruments, planning and instrumentation of weather stations.

The work of the meteorological service is dependent on a large number of individual observations. Around the world there are thousands of meteorological stations, most of which report every three hours giving details of rainfall, sunshine, windspeed and wind direction, etc. Since two-thirds of the earth's surface is covered by ocean, and observations over the ocean are very important, meteorological reports must take into account the reports from the many stations on board ship. These stations include weather ships that specialize in weather observation, and are permanently stationed at some precise point in the oceans. The weather ships release radio sondes, which provide information of conditions at altitudes of 20km and 30km (12mi and 18mi). Balloons with radio sondes also are sent up from selected land based stations to provide data twice daily.

In April 1960 the Americans put up the first weather satellite (Tiros I) into orbit. Further weather satellites have followed, from both the US and USSR. Meteorologists have been quick to appreciate the enormous importance of these new technical aids for weather forecasting. At the beginning of the 1960s the WMO formed the 'World Weather Watch', to account for the gaps in the network of the areas covered by meteorological stations, particularly over the significant equatorial and polar regions – for this, they used weather satellites. Two satellites with a relatively low orbit over both poles began a regular observation of the entire earth's surface. Four satellites were stationed at 36,000km (22,350mi) above the equator, to record weather events there. The satellites were supplemented by the establishment of further land stations and a supply of automatic buoys and weather ships, particularly in the southern hemisphere which, because there are fewer countries servicing this half of the world, is poorly represented in meteorological data.

The aim of the meteorologist is not only to improve daily weather forecasting, but also, and perhaps more particularly, of long-term and medium-term forecasts. Thirteen European countries have invested £25 million in a research centre at Reading to specialize in medium-period forecasts (up to ten days). If medium-period forecasts could achieve the accuracy of today's short-term prognoses, then the countries taking part would save yearly £250 million. The agricultural and building industries would profit most. The institute began its work in 1978.

What weather reports and forecasts are available?

Most radio stations give weather reports several times a day, usually before or after various news bulletins. These reports usually give an outline of present weather conditions and a forecast for the next 24 hours. Frequently the report is supplemented by details of observations made that morning at various stations over an extensive region, including details of pressure, temperature, wind direction and strength, and precipitation, plus perhaps reference to the state of the ground – frost, ice, snow cover, etc.

Other special meteorological services are offered at various seasons of the year, according to needs of the particular groups. Examples are weather reports and forecast for gliders, farmers, sailors, motorists (traffic reports), mountaineers (avalanche warnings) and holiday makers (travel and winter sports reports). Such special reports may be linked with news bulletins or included in special programmes aimed at the particular group. In contrast to these seasonal activities, the bulletins for shipping and coastal stations continue throughout the year.

Both national and local television networks transmit weather reports and forecasts with the news. Obviously, television has an advantage over radio in that the forecasts can be accompanied by a simplified weather map which can be designed to show the current weather picture, as well as the probable conditions for the next day. The symbols used by the BBC for Television Weather Forecasts are shown below. These

Temperature: Red figures on a yellow background give positive temperature in degrees centigrade. Black figures on a light blue background give freezing temperatures, ie. below zero centigrade.

Sunshine: The yellow symbol represents the sun; the red figures in the centre show a temperature of 25 degrees centigrade.

Cloud: A white cloud symbol indicates fine-weather clouds that may be relatively thin and patchy.

A black cloud represents the thicker and more widespread clouds often associated with dull weather.

Sunny Intervals: The sun symbol used in conjunction with a cloud in this way means some sunshine as well, particularly if the white cloud symbol is used.

Rain: The dark blue tear-drop symbols beneath the cloud indicate rain.

Rain Showers and Sunny Intervals: A combination of rain, cloud and sun represents sunny intervals and rain showers.

Snow: The white snow symbols beneath the cloud indicate snow.

Sleet: The rain and snow symbols together beneath the cloud indicate sleet.

 Thunderstorm: The symbol of a black cloud with a yellow flash represents the possibility of thunder and lightning.

 Wind Speed & Direction: The black symbol represents the wind speed and direction, the speed printed in the centre in white is in miles per hour.

FOG **Fog:** Fog is not represented by a specific symbol: it is indicated by words on the map in the general areas likely to be affected.

symbols form a basic language of weather types whose vocabulary can be extended by use of different combinations.

On all but the most simplified weather maps the symbols employed are standard (see p. 124). On many transmissions, professional meteorologists comment live on the weather conditions. Some television companies also provide special weather forecasts for particular groups. Some sports programmes, for example, give information on conditions for water sports or winter sports at the end.

In most newspapers weather reports are given often together with maps and forecasts. The maps published come from the various national meteorological services. In Britain, for example, the weather maps and the weather reports in the national daily newspapers are provided by the Meteorological Office.

The forecast map in the daily newspapers show the expected position on the day of publication – usually 12.00 GMT (Greenwich Mean Time). The forecast is up to midnight for the same day, or for the following 12 hours, with the outlook for the following two days. Information is given about temperatures by day and night, winds, cloud and precipitation. The details of the forecast may be varied to include local conditions if the paper is regional in distribution. This is not so important in the British press, but does apply on the Continent. Many newspapers, including the majority of the nationals in Britain, give simple weather reports for noon of the previous day for important foreign cities. These are mainly European but may well include New York and Montreal. A few newspapers give further details of the weather in cities overseas, including relative humidity, wind strength in miles per hour, cloud cover, duration of sunshine in hours, and rainfall for 24 hours in inches. In addition, newspapers usually provide times of sunrise and sunset, moonrise and moonset, and lighting-up time.

Weather forecasts can also be obtained by telephone and radio telephone in many European countries. In Britain, numbers to dial are given on p. 202 and in telephone directories and dialing-code books under the heading Weather Forecasts. The forecasts are regional, relating to the regional divisions in coverage within the Meteorological Office.

Automatic telephone weather service

Local area weather forecasts for many parts of the country are prepared by the Meteorological Office and recorded by the Post Office. The area telephone directories give details of the services available in the area and their telephone numbers. There is no charge apart from the telephone call.

		STD code
Bedford area (40-mile radius)	01–246 8099	
Belfast area (15-mile radius)	Belfast 8091	(0232)
Birmingham area	021–246 8091	
Bristol area (including Weston-super-Mare)	Bristol 8091	(0272)
Cardiff area	Cardiff 8091	(0222)
Devon and Cornwall	Plymouth 8091	(0752)
Edinburgh area	031–246 8091	
Essex coast	01–246 8096	
Glasgow area	041–246 8091	
Kent coast	01–246 8098	
Lancashire, Merseyside, Greater Manchester and Cheshire	061–246 8091	
Leeds, Bradford and Huddersfield area	Leeds 8091	(0532)
London area	01–246 8091	
	Guildford 8091	(0483)
Norfolk and Suffolk	Ipswich 8091	(0473)
	Norwich 8091	(0603)
North East England	Newcastle upon Tyne 8091	(0632)
N Lincs, Retford and South Humberside areas	Lincoln 8091	(0522)
N Wales coast from Conway to Chester	061–246 0893	
Notts, Leicester and Derby area	Nottingham 8091	(0602)
Sheffield area (20-mile radius)	Sheffield 8091	(0742)
S Hants and Isle of Wight (including the coastal area between Poole Harbour and Chichester)	Southampton 8091	(0703)
S W Midlands	Gloucester 8091	(0452)
Sussex coast	01–246 8097	
Thames Valley	Reading 8091	(0734)

A further service between 1 December and 30 April covers conditions in the Scottish skiing resorts. In some European countries, the meteorological office provides reports giving information to tourists or prospective tourists on the weather in other countries in Europe, northern Africa and the Middle East. The German meteorological office even provides weather information for pigeon racing, with an automatic answering service, at weekends from April onwards. In Britain weather information offices have been established at various centres (see opposite).

Weather forecasts by radio-telephone are of particular interest to shipping. The forecasts are for twelve hours, with outlook for a further twelve, and cover the sea areas within a range of several hundred miles. Warnings are given of strong winds or storm danger in the north Atlantic and North Sea.

These weather reports and forecasts distributed by the various media come from the official meteorological services in the respective countries. In Britain, the Meteorological Office provides a wide range of information and advice on flight conditions, agriculture, etc., and publish scientific papers together with statistical digests.

10°W 5° 0°

60°

Lerwick
Lerwick 2239

Kirkwall Airport
Kirkwall 2421, ext. 34

Kinloss
Forres 2161, ext. 673/674

Aberdeen Airport
Dyce 722334

Pitreavie
Inverkeithing 2566

Leuchars
Leuchars 224

GLASGOW
041-248 3451

Edinburgh Airport
031-334 7777

Prestwick
Prestwick 78475

NEWCASTLE
Newcastle 26453

55°N

Belfast Airport
Crumlin 52339

Linton-on-Ouse
Linton-on-Ouse 381

Bawtry
Doncaster 710474

MANCHESTER
061-832 6701

Binbrook
Binbrook 527

Valley
Holyhead 2288

NOTTINGHAM
0602 384092

Shawbury
Shawbury 335

Birmingham Airport
021-743 4747

Marham
Narborough
337398

Aberporth
Aberporth 810117

Wyton
Huntingdon 52451,
ext. 458

Honington
Honington 466

Brawdy
Solva 528

Gloucester
Churchdown 855566

Benson
Wallingford 36202

Lyneham
Bradenstoke 890283

LONDON
01-836 4311

Cardiff-Wales Airport
Rhoose 710343

Upavon
Upavon 286

Boscombe Down
Amesbury 3331, ext. 2131

SOUTHAMPTON
Southampton 28844

St Mawgan
Newquay 2224

Plymouth
Plymouth 42534

50°

10°W 5° 0°

International classification of clouds

The following ten photographs give the basic types of cloud. The range of forms which can be seen is, of course, much greater, but in principle all can be traced back to these basic types. A collection of colour photographs of clouds, together with explanations, published in 1967 by Pergamon Press, is *Cloud studies in colour* by R. Scorer and H. Wexler. Another book worth consulting is *Instant Weather Forecasting* by Alan Watts (published by Adlard Coles in 1968).

Cirrus (Ci): Ice clouds, feathery in appearance and belonging to the high cloud types – 5km (3.1mi) or more high. Their orientation is a key to weather development (see p. 34). Fibrous, often veil-like in appearance.

Cirrocumulus (Cu): Ice clouds, small billows, aligned small clouds, also belonging to the high cloud types – 5km (3.1mi) or more high. Significant indicators of weather change.

Altostratus (As): Grey sheets of ice and water droplets, belonging to middle altitude clouds – 2.5–7km (1.6–4.3mi) high. A significant indicator of weather change.

Cirrostratus (Cs): Ice clouds, veil-like sheets, but belonging to high cloud types – 5km (3.1mi) or more high. Usually indicate the coming of a warm front of a depression (see p. 42). Occurrence of halo phenomena (see p. 52).

Altocumulus (Ac): Mainly consisting of water, billowing clouds often arranged in bands, and belonging to middle altitude clouds − 2.5–7km (1.6–4.3mi) high. Indicative of changeable weather.

Nimbostratus (Ns): Thick sheets consisting of ice and water, belonging to middle altitude group – 2.5–7km (1.6–4.3mi) high. Forming from altostratus (see p. 86) as rain clouds, generally with marked upward billows.

Stratocumulus (Sc): Cloud sheets consisting of water droplets (cloud banks) with billows, and belonging to low cloud group – less than 2.5km (1.6mi) high. Despite threatening appearance and extent, these clouds usually bring little rain.

Stratus (St): Cloud sheets consisting of water droplets and belonging to low cloud group – less than 2.5km (1.6mi) high. Typical shapeless rain clouds often with hanging cloud trails in the mountains and near the ground.

Cumulus (Cu): Billowing clouds formed from water droplets and belonging to the low cloud group – less than 2.5km (1.6mi) high – but, when towers form, reaching great heights. When widely separated, known as fair weather cumulus (see also p. 38).

Cumulonimbus (Cb): Great billows consisting of water droplets below, ice particles at heights, belonging to the low cloud group with towers reaching as high as 10km (6.2mi), anvil-shaped, typical storm clouds; showers.

Evaluation of windspeed from water movement

(in Beaufort scale, see also p. 114)

Windspeed 0 (calm) – the remains of a swell can still be seen.

Windspeed 1 (light air) – ripples like scales but no foam crests.

Windspeed 2 (light breeze) – short small waves, no breaking of crests.

Windspeed 3 (gentle breeze) – crests gradually begin to break.

Windspeed 4 (moderate breeze) – white horses appear, waves become longer.

Windspeed 5 (fresh breeze) – white horses everywhere, long waves.

Windspeed 6 (strong breeze) – sheets of white foam formed by waves breaking; large waves develop.

Windspeed 7 (near gale) – foam sheets form in wind direction.
Windspeed 8 (gale) – foam strips in wind direction, strongly developed; spindrift from wave crests.

Windspeed 9 (strong gale) – visibility impaired by spindrift, high waves.

Windspeed 10 (storm) – sea foams: very long mountainous waves with long breaking crests; spindrift makes visibility still more difficult.

Windspeed 11 (violent storm) – breaking crests blown to froth; extraordinarily high waves, considerable loss of visibility.

Windspeed 12 (hurricane) – air filled with foam and spray.

Swell – fishing boat in a deep swell with a gentle breeze (see also p. 90).

Cross sea – stormy wind (wind sea) and, moving from left to right, a long middle-sized swell. A high cross-sea to the right of the picture.

Storm breakers – waves breaking over a sea wall on the island of Sylt.

The seaboard weather map – tips for sailors

Weather reports for shipping are indispensable for sailors on both coastal and ocean routes. For yachtsmen and inshore sailors the Meteorological Office provides plotting charts (see pp. 224–5) so that the sea area broadcasts can be seen geographically. The charts are available from Weather Centres (see p. 203) or by post from:

Southampton Weather Centre,
160 High Street,
Southampton S01 0BT.

The yachtsman can write down the general synopsis and the coastal station reports given in the Inshore Waters (up to 12 miles offshore) forecasts. These are broadcast for England and Wales at the end of Radio 4 programmes, together with the Shipping Bulletin, at 00.15hrs. The forecast for wind, weather and visibility is given first, followed by reports from Boulmer, Spurn Point, Manston, Portland Bill, Lizard, Mumbles and Valley (as marked on chart). The reports are of wind direction and force, present weather, visibility, pressure and pressure tendency. Similarly Radio Scotland gives the forecast and reports at the end of the normal broadcasting service, the reports from Machrihanish, Tiree, Stornoway, Sule Skerry, Aberdeen, Leuchars and Lerwick. Radio Ulster provides a similar service with reports from Kilkeel, Malin Head, Machrihanish, Blackpool, Valley and Orlock Head. At 06.55 hrs weekdays and 07.55 hrs weekends BBC Radio 3 gives a forecast for all inshore waters but without coastal reports.

Weather Bulletins for shipping are of two types: Atlantic weather and coastal areas. Atlantic Weather Bulletins for shipping are issued by wireless telegraphy from the Post Office radio station at Portishead. These are in six parts.

part 1. storm warnings
part 2. synopsis of weather conditions
part 3. area forecasts
part 4. analysis
part 5. N. Atlantic ship reports
part 6. land station reports

Parts 1, 2, 3, 5 and 6 are broadcast at 09.30 and 21.30 GMT, part 4 at 11.30 GMT. The information from parts 5 and 6 are plotted on board ship, using outline charts that may be purchased from the Meteorological Office. The analysis part 4, based on more information than is available to sailors, will assist in the completion of the synoptic chart.

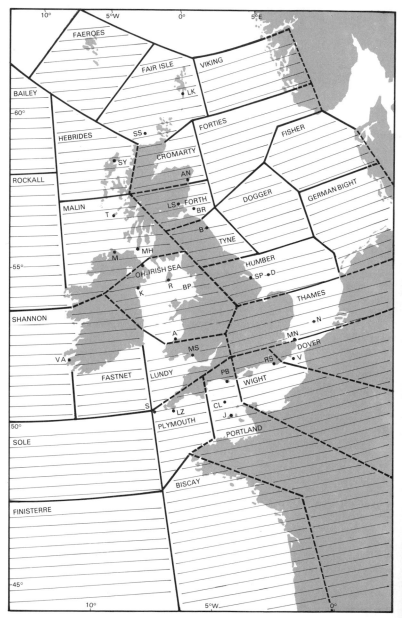

WEATHER PLOTTING CHART FOR YACHTSMEN

GENERAL SYNOPSIS at ....................G.M.T./B.S.T.............198.......

	COASTAL REPORTS at ....................G.M.T./B.S.T.	
T	TIREE	
SS	SULE SKERRY	
BR	BELL ROCK	
D	DOWSING	
N	NOORD HINDER	
V	VARNE	
RS	ROYAL SOVEREIGN	
CL	CHANNEL LIGHT VESSEL	
S	SCILLY	
VA	VALENTIA	
R	RONALDSWAY	
M	MALIN HEAD	
J	JERSEY	
B	BOULMER	
SP	SPURN POINT	
MN	MANSTON	
PB	PORTLAND BILL	
LZ	LIZARD	
MS	MUMBLES	
VY	VALLEY	
BL	BLACKPOOL	
K	KILKEEL	
MH	MACHRIHANISH	
OH	ORLOCK HEAD	
LK	LERWICK	
AN	ABERDEEN (DYCE)	
LS	LEUCHARS	

To draw in the transmitted reports on the base map, the following sequence must be observed:

1. Addition to the map of pressure systems. These are

   (a) *Secondary depressions* on the periphery of a major depression. Generally, these move quickly into the heart of the depression, but have important weather effects.

   (b) *Low pressure trough* – following a marked low pressure system, usually bringing stormy weather.

   (c) *Wave* – at first an ill-developed depression at the cold front, that can became a secondary.

   (d) *Saddle* – forming a link between two high pressure areas and separating two lows but without closed isobars.

   (e) *High pressure ridge* – an ill-developed high between two well marked highs; has a steep pressure gradient to a low pressure area.

2. Addition of fronts (cold fronts in blue, warm fronts in red, occlusions in violet). The fronts can be extremely variable in nature as shown opposite.

3. Inclusion of station reports (see p. 122) from land and sea. These begin with wind direction and speed, followed by cloud cover, observed rainfall, temperatures (shipping reports give air temperature and temperature of the surface water, air temperatures entered above a line, sea temperatures below). Obviously, the observations made on board the ship itself are entered on the seaboard weather chart.

4. Indication of isobars every 5 millibars. It is advisable to note down the values on the base map in pencil and then to add the isobars, usually in smooth curves. Concave bends in the isobars indicate low pressure areas. The isobars come to a point at fronts, evidence that the fronts indicate weather change.

   In advance of the front, the pressure usually falls more slowly than it rises after the front has passed through. The isobars in advance of the front are therefore drawn as meeting the front relatively smoothly, whereas the isobars to the rear of the front meet with it a sharp angle. *Note* that isobars do not cross or touch. In order to be accurate, it is advisable to draw in the isobars after receiving the station reports. Since the distance between the isobars and the strength of the expected wind have a definite inter-relationship, it is important for sailors not to omit any isobars on the ship's own weather map.

The weather recordings on the chart are gained from broadcasts of observations and those made on board. The official weather reports cover a period of three hours, sometimes longer, so that it is necessary to make a critical comparison with the weather observable on board. It is desirable to draw a seaboard weather map for every weather report, so that the captain has an overall view of the pressure field and fronts, paying particular attention to *direction of movement* and *speed*.
For the sailor it is important to remember that:

1. The wind veers if a depression passes to the north, and backs if a depression passes to the south.

2. The wind falls if the isobars along the ship's course draw further apart, and rises if the isobars along the course draw closer together.

The further a boat moves from the coast into open water, the more important it is to complete the seaboard weather map carefully. It would be quite foolish not to complete the map and, as a result, be blown off course and go aground.

front symbol	type of front
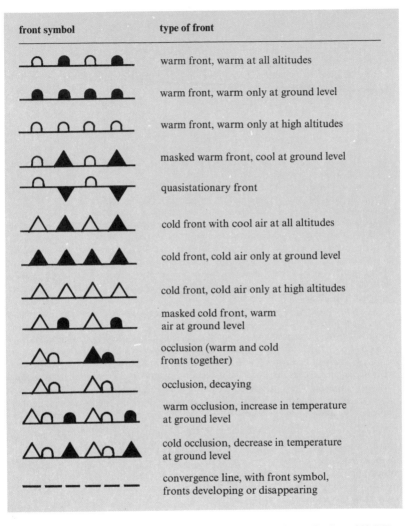	warm front, warm at all altitudes
	warm front, warm only at ground level
	warm front, warm only at high altitudes
	masked warm front, cool at ground level
	quasistationary front
	cold front with cool air at all altitudes
	cold front, cold air only at ground level
	cold front, cold air only at high altitudes
	masked cold front, warm air at ground level
	occlusion (warm and cold fronts together)
	occlusion, decaying
	warm occlusion, increase in temperature at ground level
	cold occlusion, decrease in temperature at ground level
	convergence line, with front symbol, fronts developing or disappearing

Weather reports for coastal areas are broadcast on BBC Radio 4 on 200 KHz (1500m) at the following times.

00.15 – 00.20
06.25 – 06.30   'clock' times
13.55 – 14.00
17.51 – 17.55

Each broadcast is structured as the Atlantic Weather Bulletin parts 1, 2 and 3 followed by reports from coastal stations in the order shown on the weather plotting chart. Of course it is quite possible to put the information onto an outline chart.

Gale warnings for coastal areas are broadcast by the BBC at the first programme junction. There are in addition special 24 hour forecasts for fishing fleets.

Weather forecasts at sea can be a matter of life or death so that for further study in the professional field the reader is referred to 'Meteorology for Mariners' published by the HMSO, and to 'Weather Forecasts' a Royal Yachting Association publication available from:

The Royal Yachting Association
Victoria Way, Woking, Surrey, GU21 1EQ

Weather observation

Regular observations of the weather serve both forecasting and climatology. This function is performed by the official meteorological services and observers associated with them. For others, weather observations extend the range of personal experience and assist in the understanding of nature. Many actual or would-be naturalists build themselves, as the occasion permits, a small weather observatory in the garden. Often, personal interest links with a professional interest – as with farmers, gardeners, foresters and others whose employment is weather-dependent, such as builders. Such people make use of their own observations, as well as the official weather forecasts.

Weather observations are made first with the naked eye – cloud cover, cloud type, wind direction, wind speed and certain other phenomena such as rain, storms, state of the ground, animals and plants. Those involving measurements with instruments are pressure, temperature, humidity, and amount of rainfall. Amateur observers usually only make the most straightforward observations, employing the simplest measuring apparatus. For regular recordings, as examined in the chapter on Basic Meteorology, it is best to assemble the measuring apparatus in a container for thermometers – usually called a Stevenson screen. Even if only one thermometer is used to measure air temperature, such a container or screen is necessary. The thermometer must be placed in circulating air, but in no circumstances directly in the path of the sun. It is also important to notice, when installing the equipment, that different ground surfaces have different responses to heat. Usually the container is placed above a close-cut grass sward. The screen should be free-standing and at least 10 metres (32.8ft) from buildings. During the day, temperatures at ground level on grass are higher than they are at a height. In order to obtain comparability, the meteorologist services of various countries have agreed to a standard height of 2 metres (6.56ft). It is important to ensure that the container opens northwards, so that direct sunlight does not fall on the instruments when the flap is opened. If the Stevenson screen is fully equipped, it contains:

1 dry-bulb thermometer (vertical)	1 hygrometer
1 wet-bulb thermometer (vertical)	1 thermograph
1 maximum thermometer (horizontal)	1 hygrograph
1 minimum thermometer (horizontal)	

For more about these instruments, see page 110. Since humidity is measured as well as temperature, four thermometers are always required. If you want to measure rainfall, then a rain gauge must be installed. Rainfall can be measured either as height above the surface (as if the rain cannot run off or percolate through the soil), or as the quantity received over unit area. The first is measured in millimetres or inches the second as litres or gallons per square metre or foot.

A barometer will also be needed if you wish to measure pressure (see p. 95).

In order to understand the weather, particularly to be able to compare readings from different stations, regular periodic observation is required. To be of use, the obser-

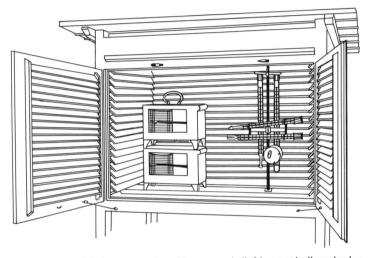

Stevenson screen with thermograph and hygrograph (left), a wet bulb and a dry bulb thermometer (right, vertical), and a maximum and a minimum thermometer (right, horizontal).

vations must be taken regularly several times a day (morning, after midday and evening), always at the same time at the same place and in the same way.

Most observers will find the weather outline sheet prepared by the meteorological services of value. This sheet can be used for recording readings over an entire month. An example of such a sheet is given on pp. 222–3 showing the recordings in July 1980 at Rogate Field Centre, Petersfield, Hants.

A valuable addition to the weather outline sheet is the recording of growth phases of important plants, showing the effects of weather on growth. The growth phases and forms of plants at different periods will mirror the weather sequence (see pp. 194 & 235).

Rogate Field Centre

Year	Mth		Wind					Screen		Screen				Soil or Earth Temperature			
80	**07**	Total Cloud	Direction	Speed	Present Weather	Visibility	Dry Bulb	Wet Bulb	Max*	Min	Grass min at 09h	Concrete min at 09h	At depth 10cm	At depth 20cm	At depth 30cm	At depth 50cm
Day	Date															
A	**B**	**C**	**D**	**E**	**F**	**G**	**H**	**I**	**J**	**K**	**L**	**M**	**N**	**O**	**P**	**Q**
T	1	7	29	09	03	7	11.4	9.8	19.0	9.2	8.5				16.0	
W	2	5	36	05	01	6	17.3	14.4	17.5	11.0	8.7				15.4	
T	3	3	14	05	01	7	17.4	15.0	20.5	6.8	5.9				15.7	
F	4	6	23	05	01	7	15.0	13.3	19.6	11.9	11.8				16.5	
S	5	4	23	03	01	7	16.8	12.5	17.5	8.6	5.3				16.1	
S	6	6	18	05	03	7	14.8	12.2	19.2	8.5	6.0				16.5	
M	7	6	23	05	01	7	14.8	13.8	17.3	11.0	11.4				16.3	
T	8	8	36	05	03	4	12.7	12.3	16.5	10.7	11.0				16.0	
W	9	8	36	05	03	6	13.2	11.9	16.2	11.0	8.8				15.5	
T	10	4	0	0	01	7	15.0	11.1	15.2	8.4	4.4				15.0	
F	11	6	0	0	01	6	15.1	12.0	16.4	5.9	4.0				15.0	
S	12	8	32	05	02	7	14.0	11.6	18.5	9.1	9.5				15.5	
S	13	4	0	0	02	7	14.5	12.4	16.7	10.5	2.0				15.5	
M	14	6	0	0	02	7	16.5	15.6	16.9	12.5	13.0				16.0	
T	15	4	34	05	01	7	14.0	11.4	17.5	9.4	7.0				15.5	
W	16	4	0	0	01	6	15.6	12.5	17.5	8.5	5.6				15.4	
T	17	6	0	0	03	6	13.3	11.0	19.0	6.5	3.7				15.6	
F	18	6	0	0	03	6	15.6	14.4	16.1	12.8	12.0				15.8	
S	19	8	23	13	03	6	15.1	14.9	16.9	13.2	13.3				16.0	
S	20	8	0	0	02	7	13.7	12.3	19.6	12.3	11.7				16.0	
M	21	4	32	05	01	6	14.8	11.8	17.8	6.8	3.3				15.6	
T	22	0	18	05	01	7	16.8	12.3	19.9	6.0	3.1				16.0	
W	23	4	0	0	03	6	16.0	14.7	20.0	6.8	5.0				16.4	
T	24	2	0	0	01	7	17.8	14.4	23.2	10.1	6.5				17.0	
F	25	0	18	05	02	6	21.1	18.3	22.1	9.4	7.5				17.5	
S	26	8	23	05	03	5	17.4	17.1	26.5	17.0	16.5				18.5	
S	27	6	23	05	03	6	16.5	15.3	19.6	13.4	10.8				17.9	
M	28	4	0	0	01	6	21.0	17.8	21.4	8.2	6.9				17.5	
T	29	6	09	05	01	4	21.0	19.2	25.4	16.4	14.0				18.5	
W	30	6	23	05	03	7	16.5	14.2	24.8	13.9	12.1				18.4	
T	31	—	—	—	—	—	—	—	—	—	—				—	

State of Ground	Snow Depth Total	Fresh	Total amount of Rain 9hr-9hr	9h–9h Run of Wind Cup-counter Reading Not Reported	9hr-9hr run of Wind	Total Sunshine for the day diff. from prev's day	Weather Diary and Remarks Before 09h GMT	After 09h GMT
S	T	U	V	W	X	Y		
2			8.7			2.1	Dull, cold	Mainly dull, cold wind
1			0.0			6.1	Brightening	Sunny, heavy showers p.m.
2			3.6			7.2	Sunny & dry	Mainly dry & sunny
2			3.4			5.7	Brightening	Dry & sunny
1			0.0			8.6	Light cloud, sunny & dry	—
1			0.0			2.0	Dull, dry	Mainly dull
2			8.6			0.0	Dull	Dull
2			15.9			0.1	Dull, showers	Dull
2			1.6			3.1	Dull	Dull
1			0.0			4.8	Light cloud, dry	Dull, dry
1			0.0			3.3	Light cloud, dry	—
1			0.0			2.7	Dull	—
1			2.2			1.6	Bright, calm	Cloudy, showers
2			8.9			2.1	Cloudy with showers	Mainly dull
			0.0			6.6	Bright, sunny	Mainly dull
			0.0			6.6	Light cloud	Bright & dry
			0.0			1.9	Dull, showers	Dull with showers
			1.2			0.0	Dull	Dull
			9.5			1.4	Rain, dull	Mainly dull
			1.7			5.1	Rain, dull	Dull
			1.0			11.5	Sunny, dry	Sunny, dry
			0.0			9.0	Sunny, dry	Sunny, dry
			0.0			8.0	Dry	Sunny & dry
			0.0			11.5	Sunny & dry	Sunny & dry
			0.0			7.8	Sunny & dry	Sunny & dry, storms early a.m.
			16.2			1.4	Stormy	Rain & sunshine
			0.6			7.2	Dull	Dry & sunny
			0.0			7.1	Dry & sunny	Dry & sunny
			0.0			2.5	Dry	Dry & sunny, storm 4 p.m.
			7.6			0.0	Dull, dark cloud	—
			—			—	—	

Gilbert White's observations of the weather

Gilbert White's observations were made from 1751 to 1773, primarily in the village of Selborne Hampshire, but with some references, for comparative purposes, to weather elsewhere. His observations were not initially supported by any measurements, and his only instrument was a mercury barometer, with later a thermometer, but they make fascinating reading. They show very clearly the value of recorded observations by amateurs, especially those with local knowledge, and are testimony to the lasting interest of such observations. Four distinctive themes of interest today can be distinguished.

Variability

A study of meteorological data, and of other evidence, shows that there have been perceptible shifts in climate over the centuries. These shifts have been of a much lesser degree than the Ice Age, although the period 1550–1850 is now considered to have been sufficiently wretched in western Europe to merit the name 'Little Ice Age'. It is known from observations supported by instrument readings that the first half of the 18th century was unusually dry, and the first 40 years of the 19th century unusually wet. Despite these minor shifts however, taken year by year the English climate still surprises with its variability, a constant goad to devoted gardeners, as White undoubtedly was. His summary of the weather of 1751 betrays his exasperation. 'The year 1751 was one of the wettest years in the memory of man. There were constant storms and gluts of rain from 20th Feb. to the 20th of May. Part of May and all of June were very dry and burning. But all July, and great part of August were as wet as ever: so that nothing in gardens in a clayey soil grew to any size: and nothing came to bear till five or six weeks later than usual'.

Memories of weather are bad or confused but as noted the first half of the 18th century was unusually dry so that White may have been correct in suggesting 1751 was the wettest in memory, but at Tonbridge, Kent, where a rain gauge was kept, 1739 was slightly wetter than 1751.

1754 showed a similar degree of variability, but in this instance towards drought. White's first entry for the year was 'an uncommon severe winter: most things in the garden destroyed'. By March 5th he was writing 'the season uncommonly dry and fine', but then on April 19th 'very thick ice and the ground frozen hard. Frequent showers of snow and hail'. By May 22nd however the weather was again 'uncommonly dry and sultry'. So it continued with the observation on July 23rd 'No kind melon-weather since the beginning of June: but a constant cloudy, windy season, but not much rain'. The drought was broken on October 9th 'a thorough soaking rain, after an uninterrupted fit of above six weeks dry weather'. White's observations are supported by the readings from the Tonbridge rain gauge with only 0.08″ in September 1754.

As remarked at the beginning of the book the only certain thing about weather is uncertainty and White's observations for subsequent years continue to show this astonishing variability of English weather. His observation for 1755 began with March

12th 'very deep snow, 7 inches on plain ground'. On July 18th he recorded 'an uncommon dry summer to this time' and on July 26th 'only six brace of melons set: hindered in their ripening by a long run of cold shady weather', similar that is to the summer of the previous year. Then on November 2nd the observation is of 'fine settled weather for 9 or 10 days before: the only good weather since July'. This fine spell occurring in late October about St Luke's Day, October 18th, is widely recognized as the 'golden October' period of German-speaking lands. In 1755 it was brought to an abrupt end for on the night of November 11–12th 'most uncommon frost for one night and considering the season of the year: ice near an inch thick, and dirt hard enough to bear an horse'. The final comment for the year was 'a terrible winter for earthquakes, inundations, tempest and continued rains. No frost worth mentioning except on the 11th and 12th of November'.

According to White 1756 was a wet year. His observations began on February 19th 'snowed hard almost all day' continued on March 23rd with 'storms of hail and rain all day' and March 24th 'hard frost and thick ice'. The cold snap continued with the recording of March 27th 'frost so fierce that it damaged the flowers and shrubs very much: ice near an inch thick: heavy snow all morning'. In April White commented 'this April thro a most surprising season for wet and frost. The 20th was a vast rain but on the 26th it rained for 22 hours without ceasing, and brought such a vast flood as has seldom been seen: the meadows around Oxon being entirely covered to a great depth'. This untoward weather continued on May 4th 'heavy storms of snow and thunder'; May 5th 'frost so hard that the dirt carried'; May 7th, 8th, and 9th 'very wet season'; May 11th 'violent rain, snow and hail: ice in the night' and May 14th 'dripping season still'. The change then was abrupt, usually a feature of continental rather than of oceanic weather, for on May 15th White wrote 'no one day so much as spring-like before; now absolute summer'. His pleasure must have been short lived, for on May 18th he recorded 'very hot sunshine with cold east wind', on May 28th 'bright sunshine and smart frosts for this fortnight past with a cutting east wind most part of the time: now a small shower' and on May 31st 'bright settled dry weather, the ground bound as hard as a stone'. This dry weather continued until June 7th when it was recorded 'three weeks and three days drought except a shower once for a few minutes. Now moderate showers'. The next weather observation relates to August 22nd 'constant heavy rains for a week: the wheat that is down begins to grow'. A further change was recorded a few days later on August 28th 'true fine harvest weather' but then as a comment on the previous rain 'wheat much grown about the country; some grew as it stood'. The frost arrived early with consecutive frosts on the nights of October 9th, 10th, 11th, and 12th, and from the 18th to 25th 'uncommon fierce frost and some snow'. White's observations again receive support from the Tonbridge gauge which shows the wettest April for the period of the records (1728–1765).

1757 appears to have been another cold dry year. A frost occurred early in January. 'On the 2nd began a frost which on the 3rd and 4th by means of a strong east wind became very severe so as to freeze up all the pools and ponds. The frost penetrated deep into the ground and seems to have been the severest since that in 1740'. On February 5th White wrote 'very dry weather and severe frost'. Wet weather ensued in April for the recording for April 2nd reads 'sad wet cold weather and constant high winds (some of them very terrible and mischievous ones) for three weeks past'. A further deterioration was reported on April 21st 'snowed very hard for sixteen hours: the greatest

snowfall that has fallen this year: and must have been a foot deep had it not for the greatest part melted as it fell. Went away without any frost'. Drought followed for on May 18th White recorded 'very hot sunny weather: no rain for five weeks; the ground very much burnt', on June 10th 'extream dry weather', and again on June 13th 'extream hot weather'. The drought broke on June 20th but was rapidly re-established. On July 5th the entry was 'drought continues and the garden suffers greatly'. The final end to the drought occurred on July 20th with 'great tempest of thunder and lightening (sic) and vast rains after 13 weeks drought. Frequent showers till the 15th of August then sixteen days wet and very bad harvest weather'. Yet a further change took place with the return of dry conditions. The entry for September 28th reads 'delicate autumn weather and no rain for more than a month'. By November 1st the drought was serious 'no rain for many weeks and the ground as dry as in the summer'. The observations for the year ended on December 9th and 26th 'very mild weather till this time' and 'very mild weather: hardly any frost yet'. White summarised the year as follows: 'The spring and summer 1757 were remarkably hot and dry. The dry weather began in passion week and continued on without any interruption (except 29th May) till the 20th of July. The air was rather cold in April and May: but the sun shining all day from a cloudless skie (sic) for many weeks dryed the ground in a very uncommon manner: and the heats of June and July quite burnt it to dust. On the twentieth of July fell a very heavy rain and extensive thunder shower after which there were moderate rains. From the 16th of August set in a very wet season of 15 or 16 days. About the beginning of September began the most delicate Autumn and lasted quite into November with very little or no frost quite to the close of the year'. Again the Tonbridge rain-gauge recordings support White's observations with a very dry June (0.29'') and a wet August (4.04''). White's journal continues to 1773, and was then replaced by the 'Naturalists' Journal'. The weather observations continued, made more precise by the use of a thermometer, until White's death in 1793, but enough of his descriptions have been quoted above to show that the present fluctuations, the occurrence of 'unseasonable' weather, is no new thing.

Passage of highs and lows

The weather associated with the passage of cyclones was first outlined systematically by Abercromby in 1883. Air-mass meteorology was developed by Norwegian meteorologists, particularly by V. Bjerknes and his son J. Bjerknes, during the first World War. Most of the science of meteorology was therefore established long after Gilbert White but nevertheless these later advances were based on the earlier patient observations of the weather by men like White, adding to the weather lore accumulated over centuries. In his recordings various weather sequences and patterns were already evident.

The association of winter high pressure with gloomy weather (due to inversions) was noted. On January 29th 1758 White made the following observation 'on this day the mercury in the weather glasses which had been mounting leasurely for many days, was got one full degree above settled fair in the parlour, and within half a degree of same in the study. My father, who has been a nice observer of that upstairs for full 37 years is certain that it never has been at that pitch before within that time. Very still, grey, close weather with wind at full east and quite a thaw tho there has been somewhat of a frost

for more than a fortnight past. Ground very dry; little rain having fallen for these three weeks past'. Similarly on January 30th 'same still gloomy weather'. The relatively warm air, high pressure, calm or gentle easterly winds and overcast skies suggest a high over the British Isles (see p. 148). The gloomy conditions were also associated with fog and with radiation frost. On January 30th 1765 White wrote 'a high barometer and N.E. wind with some dry days but frequent fogs and some sunshine'. For February 4th the entry is 'a very severe frost all day with a great rime. The ground was so very hard that the carting all day made not the least impression. There has been some frost for several days'. On February 7th 'very hard frost still, with great white dews. Things begin to freeze within (White means in the house). The sun shines hot and strong all day. The glass fell much some days agon: but rises again. It is most probable snow fell further north: here there was a scattering one morning'. Then on February 9th 'a very swift thaw' and on the 10th 'rain all day: this second frost lasted just a week'. Presumably the fall of pressure that White mentions was due to the passage of a depression probably to the south, and the heavy rain came when a second depression replaced the high pressure conditions. The high pressure with clear skies allowing radiation frost and fog all point to anticyclonic conditions and since the winds were from the north-east the anticyclone was probably situated further north than in January 1758 (see p. 144). Similar conditions appeared to have occurred in May the same year: 'harsh cold winds from north-east with an high barometer'. In summer anticyclones are generally associated with good weather and this was noted on June 19th 'hot dry weather with an high glass'.

Gilbert White also recognized the passage of lows and their associated rainbelts, although like many meteorologists often disappointed with his prognostications. So on June 21st in the same year he wrote 'picked up my hay without one drop of rain though the clouds, a sinking glass and an hollow wind threatned very hard', followed on June 22nd with 'a n. wind, a rising glass and all tokens of rain over for the present'.

Local variations

White noticed that the weather varied with topography. This was brought home to him first when he discovered that temperature in Newton Valence (the next village to the south of Selborne) was sometimes 10°F or more above that of Selborne during frosty weather. White first suspected the thermometers but brought them together to find that their readings then agreed. He discovered that Selborne, at the foot of the hanger, was in a frost pocket.

Rain due to instability showers is sharply local, in contrast that is to frontal rain which covers very wide areas. White of course noted this. Thus on June 15th 1713 he noted 'vast rain at Alton, but only a small sprinkling here. And again on July 6th 1765 'there have been fine rains round Andover and Salisbury: the verdure on the Downs is very delicate and the sheep ponds full of water. But when I came on this sides Alresford I found all the ponds without one drop of water; and the turf and corn burnt up in a very deplorable manner; and everything perishing in the garden'. And again later the same month on the 19th 'rains about but none here. Hot ripening weather. The dry fit has lasted 12 weeks yesterday. The wheat turns colour very fast'.

White's last comment above introduces the subject of **plant phenology,** the science dealing with the relationship between climate and periodic biological phenomena. In many ways the behaviour of plants with respect to the weather gives a more integrated measurement of the weather than numerous parameters. The German Meteorological Service on their monthly summary sheet (comparable with that issued by the Meteorological Office) includes phenological data. Reference has indeed already been made to this theme which is dealt with more extensively in *Seasonable Weather* by Lionel P. Smith. Gilbert White records in his journals the earliness or the lateness of harvest, and the times of flowering. Thus on January 19th 1759 White mentioned that snowdrops and some crocuses were in bloom before old December was out. The previous winter months had been very mild. In 1715 in contrast some crocuses were reported to be in bloom on January 30th. Smith shows photographs taken in January, of the same snowdrop clump for 1913 to 1942. Only in 1913 were the flowers comparable in development to 1759.

Despite the modern aids enjoyed by meteorological services, such as satellite photographs and computer analysis, the role of amateur observer such as White remains important. Without their work our weather would remain incompletely understood.

Climatic Statistics

Rainfall records started in England in 1677, at Townley near Burnley in Lancashire. From that date onwards measurements of rainfall were made for various periods and at various places, for example at Tonbridge 1728–1765, Chelmsford 1737–1742, Lambeth 1765–1782 and Somerset House 1787–1840.

Records of rainfall, temperature and pressure have been made on a more systematic and continuous basis since 1841, first at the Greenwich Observatory, later at the Kew Observatory. As already described, Gilbert White, and other amateur observers, made records of temperature and pressure as well as rainfall. We have therefore a fairly reliable picture of our climate from the late 17th century, becoming fuller, more accurate, with the progress of time. The earlier records are of course of more questionable accuracy, despite White's testing of his thermometers, but, more serious, large gaps in the network of recording stations were only slowly filled. Variations in altitude and aspect make difficult the transposing of recordings from one district to another (a difficulty which still in part exists) so that it was not until the mid-19th century that geographical aspects of our climate became clearer.

Records from before the 17th century suffer not only from this lack of geographical coverage but also from the tendency to record only the spectacular events – the droughts, floods and storms – rather than the humdrum of everyday weather. This type of 'spectacular' evidence can be supplemented to a degree from various other sources. For example the use of tree growth-rings in chronology is based upon the fact that in good seasons a tree made substantial growth, hence a thick ring, in a bad season, little growth. A similar phenological source of evidence for past climate is contained in the documents dealing with the royal forests. Charges were made for use of the forest for swine pasture, grazing on the mast and acorns. In a good year when a mild spring enabled trees to set fruit, and a good summer promoted ripening, the income to the crown was large. Thus for the forest of Alice Holt and Woolmer in north-east Hampshire, between 1245 and 1278 the range of incomes was from £17–16–3½ to 3s–3d. The table is given on page 238.

The table suggests that the weather of 1256–61 was untoward, and conversely 1246–7 and 1254–5 outstandingly good. From other sources it is known that the years 1256–58 were very wet. On the other hand other records suggest that 1252–53 were two of the driest years in history although the beech mast and acorn yield, as here valued, was good. This relationship between climate and the yield of mast and acorns is probably complicated. A severe frost at blossom time would ruin the crop whatever the subsequent weather. A drought might do little harm to the crop if the fruit were already set before the drought set in, particularly if the trees were deep-rooted in moist land.

It is nevertheless on such fragments of evidence that our knowledge of medieval climate is based, and the aim can only be to build a consistent picture. From 1841 it is quite otherwise. The amount of statistical material is huge. By 1862 some 500 stations in the British Isles were recording rainfall, and to make their measurements more

Payments for swine pasture in the forests of Alice Holt and Woolmer

1245–6	£ 1– 8– 0d
1246–7	17–16– 3½
1247–8	1– 4– 4
1248–9	8–10– 0
1249–50	7–14– 9½
1250–1	12– 0
1251–2	4–11– 0½
1252–3	1– 7– 0
1253–4	1–11– 0
1254–5	12– 2– 1
1255–6	2– 0– 4½
1256–7	7– 3
1257–8	3– 8
1258–9	7– 2
1259–60	11– 8
1260–1	3– 3
1261–2	17– 4
1262–3	13– 5
1263–4	14– 9
1264–5	2– 3
1265–6	1– 2–10
1266–7	18– 2
1267–8	7– 8
1268–9	8– 4
1269–70	3– 8
1270–1	12– 7
1271–2	8–10
1272–3	3– 6– 6
1273–4	7– 0
1274–5	16– 0
1275–6	2–12– 0
1276–7	3– 0– 0
1277–8	3–10– 0
1278–9	1– 8– 3

useful, more comprehensible, they were published annually in a digest 'British Rainfall'. This digest has subsequently been published annually and the number of recording stations, amateur and professional, has risen to over 6000. The publication gives the actual rainfall for the year for the stations, grouped in catchment areas; an example is given on pp.240–1. Since the excerpt is from 'British Rainfall 1961' measurements are still in inches.

For each of the 6000 stations is quoted an average based on 35 years recordings (when available), the actual rainfall each month, the total for the year (which can be compared with the average) and the date and amount of the maximum fall in one day. The density of observers makes possible the distinction of the geographical pattern of rainfall distribution. This is now well known in outline although even with 6000 observers subtler variations in rainfall, from one parish to another, from the weather-side to the leeside of a hill, still escape us, so that there is ample place for more amateur observers. When observation has been continuous for 35 years, average values have been established. Such averages will change a little as the 35 year period is brought forward (to prevent it becoming too distant with the passage of time). The pattern of rainfall distribution to which reference is made above is based on average values, revealing the contrast of the wetter west with a drier east, from the 74.4″ of Ambleside, the 66.9″ of Windermere to the 22.6″ of King's Lynn and the 21.6″ of Mildenhall. Against this general pattern there are the less well known variations such as the dry belt in the lee of the Welsh Mountains, with 26.7″ at Bridgnorth and 25.7″ at Shrewsbury. Another surprising variant is the wet area of southeast Hampshire and West Sussex with 39.5″ at Ditcham on the South Downs above Petersfield. With the relatively dense network of observers and a sequence of reliable results various analyses become possible, and are usually undertaken in 'British Rainfall'. For example it becomes possible to quote each month's rainfall as a percentage of the annual average in order to reveal the seasonal distribution. Also discussed and illustrated are the distributions of the number of days with rainfalls of specified quantities: .04″, .10″, and .40″ (now in mm). The maps show clearly that the heavy rainfall of the west is not the product of an endless drizzle. The west has also the highest incidence of very wet days.

Only a small number of the stations recording rainfall also record other elements. The network of stations with a full range of recordings, both amateur and professional, such as the Rogate Field Centre, is relatively open so there are substantial difficulties in transposition of data and, correspondingly, a niche for the interested amateur. Although the daily weather map has now disappeared as the result of government stringency, there is also a monthly weather report and an annual summary of monthly weather reports fulfilling something of the same function as 'British Rainfall'. The monthly summary gives temperatures, sunshine and rainfall data for a number of stations, again with reference to the average values. The temperature and sunshine records for August 1980 are given on pp.242–3, and repay study.

England and Wales Extract from British Rainfall 1961

Station	National grid reference	Altitude ft	Av. 1916-1950 in.	Jan	Feb	Mar	Apr	May	Jun	Jul	Aug	Sep	Oct	Nov	Dec	Year	Max fall in day in.	date
NORTHUMBRIAN—(contd.)																		
Gaunless (Wear)																		
Butterknowle	NZ 45 109257	630		4·91	2·15	1·31	3·50	2·17	1·53	3·73	3·23	1·83	2·91	1·70	2·77	31·74	1·59	14 Jul
West Auckland	NZ 45 177259	370		3·73	1·74	0·86	3·22	1·94	1·74	3·39	2·84	1·97	2·99	1·39	2·84	28·65	1·40	14 Jul
Bishop Auckland R.F.M. Sta.	NZ 45 215306	150		(area 35·9 sq. mi.)														
Wear																		
Willington S. Wks	NZ 45 208346	180		3·77	1·33	1·02	2·60	2·10	1·77	2·34	2·42	1·71	2·76	0·92	2·79	25·53	1·00	4 May
Sunderland Br. R.F.M. Sta.	NZ 45 264376	130		(area 254·0 sq. mi.)														
Browney (Wear)																		
Crookhall S. Wks	NZ 45 125505	700		5·11	1·62	0·96	2·18	(1·96)	2·12	4·92	3·28	1·86	3·09	1·23	3·59	31·92	1·25	14 Jul
Lanchester, Craigmore	NZ 45 165470	410	29·7	5·06	1·75	0·90	2·46	1·90	1·90	4·39	2·88	1·64	3·52	1·38	3·40	31·18	1·97	14 Jul
Ushaw College	NZ 45 219436	590	29·2	3·46	1·51	1·03	2·35	2·20	2·08	4·65	3·29	1·67	3·38	1·27	2·96	29·85	2·50	14 Jul
Burn Hall R.F.M. Sta.	NZ 45 259387	140		(area 68·9 sq. mi.)														
Wear																		
Houghall	NZ 45 283412	120	27·9	3·85	1·33	0·94	2·26	2·21	1·88	3·70	2·82	2·22	3·16	1·18	2·79	28·34	1·45	14 Jul
Durham	NZ 45 267416	340	26·3	4·06	1·11	0·92	2·35	1·90	1·87	3·67	3·22	1·59	3·01	1·04	3·15	27·89	1·73	14 Jul
Ryhope P. Sta.	NZ 45 404524	230	25·9	3·55	1·33	0·74	2·34	1·84	2·91	2·92	1·69	1·45	3·99	0·83	1·85	25·44	1·49	12 Jul
Dalton P. Sta.	NZ 45 411468	350	27·9	3·79	1·38	0·90	2·56	1·88	2·91	3·19	1·69	1·64	3·92	0·91	2·09	26·86	1·39	10 Jun
Mill Hill Res.	NZ 45 413425	510	27·7	3·86	1·47	0·95	2·41	1·80	2·67	3·28	2·80	2·04	3·69	1·06	2·40	28·43	1·33	12 Jul
New Winning P. Sta.	NZ 45 406384	400	29·2	4·06	1·49	0·95	2·56	1·75	2·52	2·67	2·04	1·97	3·68	1·12	2·95	27·76	1·31	10 Jul
Shotton Hall	NZ 45 417399	410		4·23	1·49	0·93	2·40	1·74	2·54	3·05	2·30	2·20	3·72	1·05	2·56	28·21	1·36	12 Jul
Hart Res	NZ 45 484343	170	22·9	3·41	1·16	0·84	2·27	1·90	2·10	2·43	1·99	1·89	3·71	0·93	2·25	24·88	1·42	19 Oct
West Hartlepool W.Wks.	NZ 45 508334	40	24·0	3·52	1·07	0·87	2·14	1·80	2·14	2·46	1·99	1·64	3·85	1·00	2·37	24·85	1·50	19 Oct
West Hartlepool	NZ 45 510327	30	24·5	3·31	1·03	0·83	2·08	1·60	1·94	2·31	1·79	1·45	3·71	0·89	2·30	23·24	1·54	19 Oct
West Hartlepool, Catcote Rd.	NZ 45 494299	70		3·19	1·01	0·69	1·99	1·60	1·93	2·38	1·47	1·43	3·25	1·01	2·40	22·35	1·23	12 Jul
Tees																		
Moorhouse	NY 35 758328	1840	81·2	10·14	(7·10)	4·95	4·60	3·55	3·30	7·65	8·75	5·12	11·36	6·25	6·07	78·84	2·69	16 Oct

	Grid ref.																		Date
Trout (Tees)																			
Moorhouse R.F.M. Sta.	NY 35 759336	1750		(area 4·4 sq. mi.)															
Tees																			
Forest-in-Teesdale	NY 35 872295	1250	54·9	6·47	5·78	2·88	4·01	3·12	1·98	6·48	4·47	4·55	8·11	3·82	7·01	58·68	2·25	14 Jul
Newbiggin	NY 35 915276	890	45·0	6·05	3·85	2·07	3·26	2·51	1·48	5·44	3·04	3·75	6·07	2·82	5·31	45·65	2·10	14 Jul
Dentbank R.F.M. Sta.	NY 35 932260	740		(area 83·9 sq. mi.)															
Lune (Tees)																			
Connypot	NY 35 818202	1840	60·7	5·40	4·30	3·80	4·70	2·70	3·20	8·10	5·80	5·60	7·30	3·40	5·70	60·00		
Dow Crag	NY 35 838192	1720	60·9	6·70	4·50	4·10	5·30	2·60	3·50	7·20	5·60	5·30	7·50	3·40	6·50	62·20		
Keekham	NY 35 823222	1660	67·5	7·80	5·50	4·50	5·00	3·30	3·40	9·00	(6·20)	6·20	8·30	4·70	6·70	(70·60)	3·10	14 Jul
Fish Lake	NY 35 851228	1500		5·59	4·92	3·97	4·50	3·32	3·31	8·43	6·20	6·07	8·47	4·63	6·40	65·81		
Lunehead	NY 35 850215	1450	60·2	7·00	4·70	3·70	4·90	2·80	3·00	7·20	5·60	4·80	7·90	3·70	6·70	62·00		
Hargill	NY 35 879222	1320	57·3	6·40	5·40	4·20	4·50	3·10	2·80	6·60	5·10	5·00	7·60	4·50	5·70	60·90		
Selset	NY 35 918217	1080	42·4	4·80	3·50	2·75	3·80	2·20	2·20	5·30	4·20	2·50	5·90	3·60	4·00	44·55		
Grassholme Res.	NY 35 946224	940	41·0	5·54	3·83	2·14	3·78	2·48	1·53	5·17	3·69	2·64	5·66	2·95	4·87	44·28	1·99	14 Jul
Tees																			
Eggleston	NZ 45 011229	1010	34·6	5·07	2·49	1·43	3·42	2·11	1·36	4·84	2·57	1·94	3·82	2·53	3·90	35·48	1·86	14 Jul
Balder (Tees)																			
Aygill	NY 35 886169	1250	49·7	5·50	4·30	3·20	4·20	2·40	2·50	5·80	5·00	3·00	6·80	3·00	(5·10)	(50·80)		
Balderhead	NY 35 886178	1360	50·6	6·10	3·90	3·10	4·00	2·40	2·50	5·80	5·00	3·00	6·90	3·20	5·00	50·90		
Blackton Res.	NY 35 930182	940	43·8	5·40	3·70	2·56	3·27	2·72	2·13	5·11	3·93	2·98	6·24	3·70	4·44	46·18		
Hury Res.	NY 35 967192	860	38·0	5·06	3·07	1·89	3·65	2·43	1·59	4·72	3·12	2·34	5·17	2·69	4·13	39·86	1·36	14 Jul
Tees																			
Lartington Filters	NZ 45 012180	730	32·8	4·86	2·72	1·45	3·77	2·18	1·32	4·78	2·68	1·65	4·12	2·43	3·67	35·63	1·40	12 Jul
Barnard Castle, Bowes Mus.	NZ 45 056164	560	31·5	4·38	2·26	1·08	3·74	1·80	1·24	3·18	2·05	1·42	3·39	1·96	3·27	29·77	1·15	15 Jul
Geta (Tees)																			
Rutherford Br. R.F.M. Sta.	NZ 45 034122	730		(area 33·3 sq. mi.)															
Tees																			
Raby Castle	NZ 45 128221	460	30·3	4·60	1·94	1·48	3·31	1·85	1·35	3·24	2·40	1·60	2·66	1·31	3·31	29·05	1·33	14 Jul
Darlington, Broken Scar	NZ 45 256141	160	26·3	3·81	1·24	0·66	3·31	1·58	1·96	2·00	1·80	1·71	3·04	0·98	2·65	24·74	1·35	10 Jun
Darlington, W. Works.	NZ 45 258139	160	25·2	3·72	1·18	0·67	3·26	1·57	2·11	1·99	1·76	1·71	3·01	0·90	2·67	24·55	1·46	10 Jun
Broken Scar R.F.M. Sta.	NZ 45 259137	120		(area 83·9 sq. mi.)															

241

Monthly summary: August 1980

Place	Temperature								No. of days with grass minimum below 0° C
	Mean maximum °C	Difference from average	Mean minimum °C	Difference from average	Highest maximum °C	Date	Lowest minimum °C	Date	
Kew	21.3	+0.4	12.5	+0.5	24.6	02	04.2	25	03
Boscombe Down	20.4	−0.1	11.6	+0.4	24.0	13	05.3	23	01
Rhoose	19.3	−0.8	12.4	+0.1	22.9	27	05.7	24	00
Mount Batten	18.8	−0.2	13.2	+0.5	21.8	02	06.7	24	00
Elmdon	19.9	−0.1	11.3	+0.4	25.1	20	03.7	25	01
Valley	17.7	−0.6	12.8	+0.6	21.7	26	05.8	24	00
Ringway	19.0	−0.4	12.6	+1.1	22.9	*	06.7	25	00
Watnall	19.4	−0.5	11.7	+0.6	23.1	13	07.2	24	00
Finningley	20.2	−0.4	11.4	0.0	24.4	03	05.7	25	00
Tynemouth	17.0	−0.2	12.3	+0.3	21.0	14	08.4	23	00
Eskdalemuir	16.7	−0.4	09.0	+0.2	21.0	26	01.7	31	04
Abbotsinch	17.9	−0.6	10.9	+0.3	21.0	14	02.0	23	03
Leuchars	18.2	+0.3	10.5	+0.1	21.2	28	03.3	23	02
Dyce	17.3	+0.2	10.9	+1.5	22.9	14	05.8	02	00
Stornoway	15.4	−0.4	11.0	+1.0	18.7	16	05.7	31	00
Aldergrove	18.5	+0.5	11.5	+1.0	21.5	13	04.6	24	01

*02,14

The written text of the monthly summary (for August 1980) observes that temperature and sunshine totals were generally below average. It was, perhaps it is best forgotten, an awful summer. This appraisal is borne out in the tables with sunshine totals at Abbotsinch (Glasgow), Valley and Aldergrove only about three-quarters of average. Kew nearly reached average sunshine values but far below the record sunny August of 1947 with 279 hours. The monthly report further illustrates the data with graphs.

Sunshine

Days of no sunshine	Maximum duration	Date	Total for month	% of average	First year of record	Highest and lowest totals on record for month			
						Highest	Year	Lowest	Year
	hrs.		hrs.			hrs.		hrs.	
02	12.4	31	176	096	1880	279	1947	109	1912
01	11.1	16	185	104	1933	298	1976	119	1968
03	12.0	21	156	087	1954	299	1976	133	1958
04	13.3	19	167	087	1921	308	1976	114	1958
00	11.0	26	150	092	1945	258	1947	104	1946
04	13.0	24	130	071	1946	291	1976	123	1948
04	11.0	25	129	085	1945	259	1947	089	1948
03	12.0	31	133	089	1948	224	1975	098	1963
04	12.3	16	130	084	1958	213	1976	112	1963
02	12.1	21	132	091	1936	247	1976	090	1942
06	13.2	23	100	078	1911	260	1947	044	1912
03	13.1	23	109	076	1921	222	1947	085	1942
02	10.9	22	129	085	1922	218	1976	097	1963
05	10.8	02	132	096	1941	257	1947	076	1963
05	10.8	30	110	082	1880	262	1899	075	1942
02	10.2	23	105	076	1927	276	1976	082	1942

These are for temperatures, sunshine and rainfall for three stations: Kew, Manchester and Glasgow, rainfall of course now measured in millimetres, and temperatures only in centigrade. Graphs for sunshine and rainfall at Manchester and Glasgow are given on p.244. The sunshine graph shows clearly the dreary first three weeks of August. Rainfall in Glasgow was almost 50% above normal. Manchester achieved its average rainfall.

Sunshine Graph

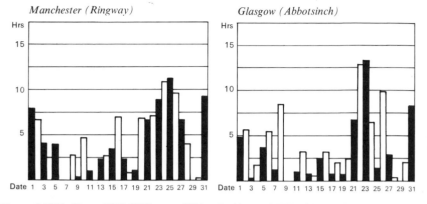

Manchester (Ringway)

Glasgow (Abbotsinch)

Total for month 129 hrs 30-year (1941–1970) average 152 hrs. Total for month 109 hrs 30-year (1941–1970) average 143 h

Rainfall Graph

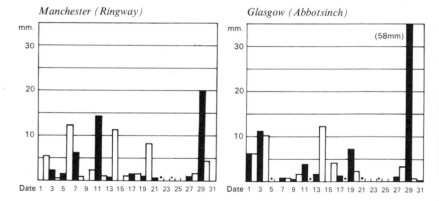

Manchester (Ringway)

Glasgow (Abbotsinch)

Total for month 91 mm 30-year (1941–1970) average 91 mm. Total for month 129 mm 30-year (1941–1970) average 89

Glossary of some basic meteorological terms

Adiabatic process. Process by which air expands or is compressed without gaining or losing heat.

Altimeter. Instrument for measuring the altitude or height above a specified reference level (sea or ground). Usually it is an aneroid barometer (see barometer).

Anticyclone. An atmospheric condition of high barometric pressure. Extending over a wide area, moves slowly and generally eastwards in the temperate zones.

Atmosphere. The gaseous envelope of the earth, in the underlayers of which – up to 15km (9.3mi) high – weather events take place. It consists of a mixture of nitrogen (78%), oxygen (21%) and mixed gases (1%), which include carbon dioxide, so important in climate.

Azores high. High pressure area in the vicinity of the Azores. This belongs to the subtropical high pressure zone of the northern hemisphere. It is often associated with the establishment of good weather in western and central Europe, but also has a role (together with the Icelandic low) in directing the inflow of westerly air.

Barometer. Instrument for measuring atmospheric pressure at a specified reference level or the height above sea (or ground) level. The mercury barometer is an evacuated and calibrated glass tube, in which a column of mercury rises or falls as pressure of the atmosphere increases or decreases. The aneroid barometer consists of a needle connected to the top of a metal box in which a partial vacuum is maintained. A change in atmospheric pressure moves the needle by forcing the box top to bend in or out.

Beaufort scale. A scale for measuring windspeed. It ranges from 0 to 12 and covers windspeeds up to 118km/hr (73mph), describing states from calm to storm. Hurricane force is in excess of 12 on the scale.

Blackthorn winter (Ice saints). Period of cold weather and late frost, the cause being a northerly flow with cold-air inbreaks, typical of the early year and early summer. On the Continent, often called 'Ice saints' winter after the saints (Pancras, Servatius, Boniface and 'cold' Sophia), whose days are 12–15 May.

Clouds. Formed by the cooling of air at high altitudes leads to condensation of water vapour. The liquid water can fall as rain, or re-evaporate into the air – the clouds may disappear without rain. Most common forms are cumulus and stratus clouds.

Continental high. Stationary high pressure area over eastern Europe (USSR). Southeasterly winds bring warm air to western, central and northern Europe. Characterized by long periods of good weather in summer and autumn.

Coriolis force. An inertial force caused by the earth's rotation which deflects moving air. The air moves to the right in the northern hemisphere and to the left in the southern hemisphere.

Corona. Optical phenomena of the atmosphere caused by refraction of sunlight or moonlight via cloud water droplets. A yellow-white circle forms around the sun or moon. Intensive development of coronae indicates building of high stratus clouds and poor weather.

Cyclone. See **Low.**

Damming-back of air streams. Air masses are dammed back by mountain chains when they approach such chains at right angles, e.g., cold air from the north of the Alps. Often associated with long periods of rain.

Depression. See **Low.**

Dog days. The end of July and beginning of August, popularly associated with high summer, hot summer days. This depends on the establishment of a stationary high pressure system in the second half of July. The name is derived from Sirius, the dog star visible in the morning skies.

Exosphere. The outermost portion of the earth's atmosphere, lying beyond the stratosphere.

Fall winds. Formed by mountains which block or deflect the movement of air. Most famous are the Föhn winds drawn up on the southern side of the Alps and warmed by descent on the north. Föhn winds can bring about a rise of temperature of 10°–15°C in a few hours. The air is very clear, with typical clouds. Other examples include the mistral and sirocco.

Front. The boundary between two masses of air that differ in density or temperature. A warm front is one along which warmer air replaces colder air, whereas a cold front is one along which colder air replaces warmer air.

Genoa low. A depression in the western Mediterranean. Cold polar air moving across the Alps generates low pressure systems, linked with unstable weather in northern Italy and the French Riviera.

'Greenhouse' effect. The retention of heat from sunlight at the earth's surface. It is caused by carbon dioxide absorbing some of the long-wave radiation emitted by the earth.

Hail. Form of precipitation consisting of ice particles more than 1cm (0.39in) thick – in fact, as large as hen's eggs are possible. The grains grow slowly as ice is deposited in its movement through the layers of air. Hail is usually associated with storms.

Halo. Optical phenomenon in the atmosphere consisting of a ring at 22° or 46° around the sun or moon. Caused by refraction of sun or moonlight, through ice crystals. The ice crystals are present in cirrus clouds. The halo is therefore important as an indicator of bad weather.

Haze. Agglomeration of water vapour molecules on condensation nucleii. An initial stage in the development of precipitation. Haze reduces the visibility but, if visibility falls beneath 1000 metres (3280ft), conditions are described as mist.

High. An area of high pressure. Subsiding air movement has the effect of absorbing cloud, bringing fine weather. A zone of permanent high pressure is present 20°–30° north (subtropical high pressure zone). Wedges of high pressure have only short-term effects. See also **Anticyclones.**

Hoarfrost (rime). Precipitation in solid form. It occurs when the temperature falls below freezing point and water vapour is sublimated, moving directly into a solid form, on the plants or soil as a white smooth coating. On roads, it leads to problems of skidding.

Ice. Crystallized water caused by cooling to 0°C (freezing point). It occurs in various forms: snow, hail, glacier ice, hoarfrost, black ice.

Icelandic low. Low pressure system over the northern Atlantic centred near Iceland. Throughout the year and, particularly in conjunction with the Azores high, this influences the characteristic westerly situation for western, northern and central Europe, bringing wind and rain, with short periods of better weather.

Indian summer. Good weather in late summer and early autumn (September) caused by high pressure over eastern North America. It can last several weeks. Similar conditions can occur during October in west and central Europe; known in the latter as 'Old Wives' summer'.

Inversion. Condition in atmosphere in which layer of warm air traps cooler air near the surface of the earth (ground level inversion). This prevents normal vertical air movements, and leads to the accumulation of industrial exhaust gases beneath the warm air. This is particularly marked in autumn and winter over cities and built-up areas.

Ionosphere. The outer part of the earth's atmosphere, beginning at an altitude of about 40km (25mi) and extending to the highest part (exosphere) of the atmosphere. It contains a series of constantly-changing layers. These layers are characterized by a significant electron and ion content.

Isobar. The line on a map which connects points on the earth's surface having equal barometric pressure at a given time or over a given period of time.

Low. An area of low pressure. Vortex-form movement of air masses, with the formation of warm and cold fronts. Cloud-building processes predominate. It brings important weather features, notably wind and rain. Typical of westerly and northerly conditions. A series of lows in sequence is typical of westerly weather conditions, occurring in Europe at any time of year. The weather is changeable with much rain and wind, interspersed with short clearances.

Major weather system. An extensive weather sequence with a tendency to recur seasonally or annually. Dependent, for example, on radiation. Example: continental high over eastern Europe in late summer. (Important for long-term weather forecasts.)

Meteorologically significant days. These are days that, according to traditional lore, mark the beginning of a long-term weather type. They are based on experience that cold and warm air masses are typical for certain seasons. Examples are St Swithin's day and the 'Ice saints' of central European calendar.

Mist. An increase in the number of water droplets in the air affecting visibility. Predominantly caused by a mixing of cold and warm air. There are ground mists and hill mists, often associated with fine penetrating rain.

Monsoon. Seasonal wind blowing in summer from a cool sea towards a warm land (summer monsoon), and in winter blowing from a cold land to a warmer sea (winter monsoon). Monsoons are typical of Asia, but they also occur in Europe.

Occlusion. Final stage of depression reached when the warm front is overtaken by the cold front. The original air mass is forced aloft – along the surface of the cold front.

Polar air. Cold air originating in northern Europe and in the polar regions. Cold air pushes warm air upwards, a process leading to cloud growth and rainfall (cold front). Maritime polar air is warmed by its passage over the Atlantic, whereas continental polar air is particularly cold.

Pressure, Air. The pressure of the atmosphere at a given place on the earth's surface. It is measured by a mercury column (760mm–29.92in–at sea level) or in millibars (1013.2mb at sea level). Air pressure diminishes with increasing altitude. The distribution of differences in pressure is an essential part of weather forecasting.

Psychrometer. An instrument for measuring the amount of moisture in the air (humidity). It consists of a wet- and a dry-bulb thermometer.

Radiant energy. Any form of energy which travels in waves, notably heat and light. Also, x-rays, gamma rays, etc.

Rain. Form of precipitation measuring up to 6mm (0.24in) diameter. Rain drops develop by the increasing collision of water droplets and ice particles, until they are heavy enough to fall. Very fine droplets form drizzle.

Rainbow. Optical phenomenon of the atmosphere formed when sunlight falls on the side of a shower cloud, such as when rain clouds are clearing. The light is refracted displaying the spectral colours from red to violet. The larger the raindrops, the more strongly the rainbow is developed.

Sensitivity to weather. Animals have a more immediate sensitivity than men to weather, and their behaviour is often an indicator of weather change. Animals do not, however, react to imminent weather, only to processes already present in the area.

'Sheep' cold period. The first half of June that, according to tradition, is associated with cold rainy days – lasting to the middle of the month. It is caused by a seasonal northerly situation with an inbreak of cold air, and usually lasts two days.

Sleet. Form of precipitation consisting of small opaque ice crystals, formed by impact of super-cooled water droplets with ice crystals, mixed with rain or snow. Droplets and crystals freeze together. Usually associated with storms.

Snow. Solid precipitation occurring while the temperature is beneath 0°C. Appears as six-pointed ice stars. Beautiful small ice crystals develop at great heights at −12°C to −16°C. Growth of snowflakes is particularly marked at about 0°C and, at this temperature, there are heavy falls.

Stratosphere. The atmospheric zone above the troposphere. It extends upwards from about 10km (6mi) to 25km (15mi) above the earth's surface. The temperature is fairly constant ranging generally from −45°C to −75°C.

Temperature, Air. The temperature of the air decreases with altitude up to 10km (6.2mi). It then remains constant for some distance, and subsequently increases gradually at further altitudes. Air temperature controls the initiating mechanism for weather events (cold and warm air masses). It influences air pressure and humidity.

Thermal wind. Rising warm air, caused by an uneven heating of the ground or sea by the sun.

Thermometer. Instrument for measuring temperature, consisting of a calibrated glass tube with a sealed, very slender bore. The bore contains mercury or coloured alcohol, which rises or falls as it expands or contracts as a reaction to changes in temperature.

Thunderstorm. Consequence of rapidly-rising moist warm air, reaching 10km (6.2mi) in height. Towering cumulus are formed with heavy rainfall (rain, sleet, hail). Electrical discharges take place, and with large potential differences – lightning.

Trade wind. Air movement blowing steadily towards the equator – from the northeast in the tropics north of the equator and from the southeast in the tropics south of the equator. Long identified by sailors.

Tropopause. The transition zone between the troposphere and the stratosphere. At this stage, the drop in temperature with increasing height ceases.

Troposphere. The atmosphere from the earth's surface to the tropopause. It extends upwards to about 10km (6mi) at the equator and 20km (12mi) at the poles. Clouds form in this stratum and temperature usually decreases with increasing altitude.

Trough. A particularly intensive low pressure system behind a cold front in a cyclone. Associated with stormy winds and strong rain. Typical around Iceland, but also occurring in the Mediterranean (e.g. in summer, with flood threat in lowlands, snow in the mountains).

Warm air. Air originating in southern latitudes, particularly in the subtropical high pressure zone. Warm air masses move up over cold air masses, resulting in cloud and rain (warm front). Typical of warm air are wide rain belts, but the weather is usually less disturbed than that associated with cold air inbreaks.

Wedge of high pressure. Moving high pressure zone between two depressions, one following upon the other. The improvement in weather usually lasts only for one or two days. The high does not extend vertically to any great degree, the air movement at height normally being a pure westerly flow.

Wind. The process by which the difference of pressure between two places is equalized. The air stream is from places with a high pressure to places with a low pressure. Wind strength reaches storm and hurricane (Beaufort scale). Wind, pressure, and temperature are functionally related to weather events.

Index

251

sparse

very sparse